HOME LINKS

Everyday
Mathematics®

The University of Chicago School Mathematics Project

Mc
Graw
Hill
Education

The University of Chicago School Mathematics Project

Max Bell, Director, *Everyday Mathematics* First Edition; James McBride, Director, *Everyday Mathematics* Second Edition; Andy Isaacs, Director, *Everyday Mathematics* Third, CCSS, and Fourth Editions; Amy Dillard, Associate Director, *Everyday Mathematics* Third Edition; Rachel Malpass McCall, Associate Director, *Everyday Mathematics* CCSS and Fourth Editions; Mary Ellen Dairyko, Associate Director, *Everyday Mathematics* Fourth Edition

Authors
Max Bell, John Bretzlauf, Amy Dillard, Robert Hartfield, Andy Isaacs, James McBride, Kathleen Pitvorec, Peter Saecker, ‡Sarah R. Burns, *Ann McCarty, Robert Balfanz, †William Carroll

*Third Edition only
†First Edition only
‡Common Core State Standards Edition only

Fourth Edition Grade 6 Team Leader
Kathleen Pitvorec

Writers
Jorge Abner Bardeguez-Delgado, Kelly Darke, Cathy Hynes Feldman, Jennifer L. Jankowski, Soundarya Radhakrishnan

Differentiation Team
Ava Belisle-Chatterjee, Leader; Jean Capper, Martin Gartzman, Barbara Molina

Digital Development Team
Carla Agard-Strickland, Leader; John Benson, Gregory Berns-Leone, Juan Camilo Acevedo

Virtual Learning Community
Meg Schleppenbach Bates, Cheryl G. Moran, Margaret Sharkey

Technical Art
Diana Barrie, Senior Artist; Cherry Inthalangsy

UCSMP Editorial
Lila K.S. Goldstein, Senior Editor; Serena Hohmann, Rachel Jacobs, Kristen Pasmore, Delna Weil

Field Test Coordination
Denise A. Porter, Angela Schieffer, Amanda L. Zimolzak

Field Test Teachers
Jason Antesbeger, Kristin M. Arras, Catherine Ditto, Benjamin Kovacs, Joshua Ryan Marburger, Dawn A. Meziere, Kyle Radcliff, Lauren Scherer, Sara Sharp

Digital Field Test Teachers
Colleen Girard, Michelle Kutanovski, Gina Cipriani, Retonyar Ringold, Catherine Rollings, Julia Schacht, Christine Molina-Rebecca, Monica Diaz de Leon, Tiffany Barnes, Andrea Bonanno-Lersch, Debra Fields, Kellie Johnson, Elyse D'Andrea, Katie Fielden, Jamie Henry, Jill Parisi, Lauren Wolkhamer, Kenecia Moore, Julie Spaite, Sue White, Damaris Miles, Kelly Fitzgerald

Contributors
John Benson, Kelley E. Buchheister, Kathryn B. Chval, Andy Carter, James Flanders, Lila K.S. Goldstein, Aaron T. Hill, Serena Hohmann, Jeanne Mills DiDomenico, Denise Porter, Kathryn M. Rich, Mollie Rudnick, Sheila Sconiers, Laurie K. Thrasher, Penny Williams

Center for Elementary Mathematics and Science Education Administration
Martin Gartzman, Executive Director; Meri B. Fohran, Jose J. Fragoso, Jr., Regina Littleton, Laurie K. Thrasher

External Reviewers
The *Everyday Mathematics* authors gratefully acknowledge the work of the many scholars and teachers who reviewed plans for this edition. All decisions regarding the content and pedagogy of *Everyday Mathematics* were made by the authors and do not necessarily reflect the views of those listed below.

Elizabeth Babcock, California Academy of Sciences; Arthur J. Baroody, University of Illinois at Urbana-Champaign and University of Denver; Dawn Berk, University of Delaware; Diane J. Briars, Pittsburgh, Pennsylvania; Kathryn B. Chval, University of Missouri–Columbia; Kathleen Cramer, University of Minnesota; Ethan Danahy, Tufts University; Tom de Boor, Grunwald Associates; Louis V. DiBello, University of Illinois at Chicago; Corey Drake, Michigan State University; David Foster, Silicon Valley Mathematics Initiative; Funda Gönülateş, Michigan State University; M. Kathleen Heid, Pennsylvania State University; Natalie Jakucyn, Glenbrook South High School, Glenview, IL; Richard G. Kron, University of Chicago; Richard Lehrer, Vanderbilt University; Susan C. Levine, University of Chicago; Lorraine M. Males, University of Nebraska-Lincoln; Dr. George Mehler, Temple University and Central Bucks School District, Pennsylvania; Kenny Huy Nguyen, North Carolina State University; Mark Oreglia, University of Chicago; Sandra Overcash, Virginia Beach City Public Schools, Virginia; Raedy M. Ping, University of Chicago; Kevin L. Polk, Aveniros LLC; Sarah R. Powell, University of Texas at Austin; Janine T. Remillard, University of Pennsylvania; John P. Smith III, Michigan State University; Mary Kay Stein, University of Pittsburgh; Dale Truding, Arlington Heights District 25, Arlington Heights, Illinois; Judith S. Zawojewski, Illinois Institute of Technology

Note
Many people have contributed to the creation of *Everyday Mathematics*. Visit http://everydaymath.uchicago.edu/authors/ for biographical sketches of *Everyday Mathematics* Fourth Edition staff and copyright pages from earlier editions.

www.everydaymath.com

Send all inquiries to:
McGraw-Hill Education
8787 Orion Place
Columbus, OH 43240

ISBN: 978-0-02-140795-8
MHID: 0-02-140795-9

Printed in the United States of America.

2 3 4 5 6 7 8 9 QVS 20 19 18 17 16 15

Contents

Unit 8

Introduction to *Sixth Grade Everyday Mathematics*

Everyday Mathematics offers students a broad mathematical background that relies on the latest research and field-test experience. This year students will be using:

- a problem-solving approach to find solutions for problems in everyday situations
- activities to develop concepts and skills as well as confidence
- repeated review to revisit concepts throughout the school year
- communication skills to explore mathematical ideas
- games as an alternative for drills to practice concepts

Content from the Common Core State Standards

Ratios and Proportional Reasoning
- Working with ratios and rates to solve problems
- Finding percent of a quantity as a rate per 100
- Using ratio reasoning to convert measurement units

Number Systems
- Dividing fractions; and adding, subtracting, multiplying, and dividing multidigit decimals
- Finding greatest common factors and least common multiples for pairs of numbers
- Comparing and ordering positive and negative rational numbers
- Naming and plotting rational numbers on a number line and coordinate grid
- Recognizing absolute value as distances on a number line

Expressions and Equations
- Writing numeric and algebraic expressions, and identifying equivalent expressions
- Creating and extending numerical patterns
- Using variables to represent numbers and write expressions when solving problems
- Solving equations and inequalities
- Applying algebraic properties

Geometry
- Calculating polygon areas and the surface area and volume of 3-dimensional shapes
- Drawing polygons on the coordinate plane

Statistics and Probability
- Collecting, organizing, displaying, and analyzing data
- Describing and analyzing patterns in data focused on variability and measures of center

This year you will receive family letters describing each unit, listing vocabulary definitions, and suggesting at-home activities. Parents and guardians are encouraged to share ideas pertaining to these math concepts with their children in their home language.

Data Displays and Number Systems

Unit 1 introduces reading, interpreting, and analyzing data. We are surrounded with data in our daily lives. News is often accompanied by charts and graphs. Businesses analyze and display data to work efficiently. Citizens must interpret data to participate in democracy. Consumers interpret and question data to make informed choices.

In *Everyday Mathematics* data provide contexts for the practice of numerical skills. In Unit 1 your child works with tables, bar graphs, histograms, and dot plots.

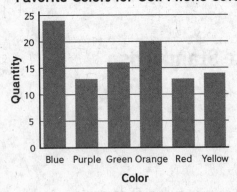
Favorite Colors for Cell Phone Covers

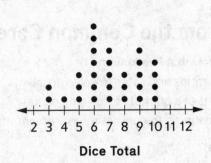
Results for 30 Rolls of 2 Dice

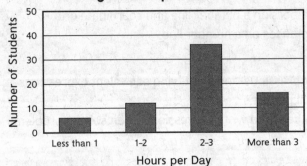
Average Time Spent on Homework

Throughout Unit 1, students think critically about representations of real-world data. Students consider the following questions:

- What is the purpose of the graph or table? Can the display be improved for clarity?
- Does the display seem accurate, or is it biased?
- Where is the center of the data and how is the data distributed?
- Can you draw conclusions or make predictions based on the data display?

Unit 1 also expands students' experience with number systems, including work with fractions and negative numbers. Students identify, categorize, and order rational numbers and use them in real-life situations. For example, students investigate how 0 and positive and negative numbers are used in contexts like temperature and elevation. Finally students plot positive and negative numbers on a number line and on a 4-quadrant grid.

Please keep this Family Letter for reference as your child works through Unit 1.

Vocabulary

coordinate grid A grid formed by two number lines intersecting at 0 to form right angles. Ordered pairs of numbers called *coordinates* locate points on the grid.

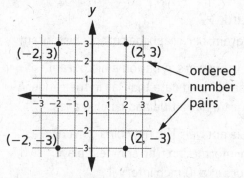

counting numbers Numbers typically used to count things {1, 2, 3, 4, . . . }.

data Information that is gathered by counting, measuring, questioning, or observing.

dot plot A data sketch in which dots show frequencies of values above a number line.

histogram A bar graph of numerical data grouped into bins or intervals along an axis.

integers Whole numbers and their opposites { . . . , −3, −2, −1, 0, 1, 2, 3, . . . }.

landmark A notable feature of a data set. Some landmarks are the median, mode, maximum, minimum, and range.

measure of center A value representing what is typical or central to a data set, such as mean or median.

measure of spread A value representing the variability of a data set, such as range or interquartile range (IQR).

rational numbers Numbers that can be named as fractions, such as $\frac{2}{3}$, 60% $\left(\frac{3}{5}\right)$, and −1.25 $\left(-\frac{5}{4}\right)$.

whole numbers Counting numbers and 0 {0, 1, 2, 3, . . . }.

Do-Anytime Activities

1. Encourage your child to use math vocabulary. Help your child recognize everyday uses of fractions, decimals, and negative numbers in media, science, statistics, business, sports, and so on.

2. Help your child find examples of graphs in newspapers or magazines to bring to class for a Graph Museum. Discuss the features of any graphs you find before sending them to school.

3. Encourage your child to practice mental math skills in everyday situations like figuring change, calculating a tip, estimating the number of people in a room, and so on.

Building Skills through Games

Games are integral to the *Everyday Mathematics* program as an effective and interactive way to practice skills. Game directions can be found in the *Student Reference Book*.

In this unit your child will work on a variety of skills by playing the following games:

Factor Captor Students find factors.

Build-It Students compare and order fractions.

Fraction Capture Students find equivalent fractions and add with common denominators.

Divisibility Dash Students divide numbers.

Mixed-Number Spin Students practice fraction and mixed number addition and subtraction.

Landmark Shark Students find mean, median, mode(s), and range for a set of numbers.

Hidden Treasure Students plot points on a 4-quadrant grid.

As You Help Your Child with Homework

As your child brings assignments home, you might want to go over the instructions together, clarifying as necessary. The answers listed below will guide you through the unit's Home Links.

Home Link 1-2

2. 90

6. Sample answer: Allowance; Unit: Dollars

Home Link 1-3

1. 21   2. Median: 23; Mode: 20; Mean: 24

Home Link 1-4

1. Sample answer: It's not likely that everyone watched exactly 4 movies.

2. 6 movies; Sample answer: The dot at 2 is two to the left of 4. In order to balance out the 2, I placed the dot 2 units to the right of 4, at 6.

3. The answer can be any combination that totals 12; Sample answer: Two 6s would balance the two 2s around 4.

Home Link 1-5

1. Mia: Median 80; Mean 82
   Nico: Median: 80; Mean: 76

2. Sample answers:
   Mia: The median, because she only has 3 scores above 80, and the mean is above 80.
   Nico: The median, because his two low scores make it look like he is doing worse than he is.

Home Link 1-6

2. Sample answer: The scales for the vertical axes are different.

3. Sample answer: Graph B will be more convincing because it is a more dramatic representation of the data.

4. Sample answer: Before it looked like there was a strong preference for tacos, but now they all appear to be close together.

Home Link 1-7

1. Histogram; bar graph; bar graph; histogram

3. Sample answers: A bar on a histogram includes binned data that fall in the range. A histogram has a title and labeled axes.

7. Sample answer: The new bins don't give as much information. The shortest player is with 3 others in a 10-inch interval.

Home Link 1-8

2. Sample answer: Maybe something different about the track made the times slower.

3. Sample answer: The mean is greater than the median and mode because of the small peak.

Home Link 1-9

3a. 2,400 feet   4a. About 2,600–2,700 miles

Home Link 1-10

3. $2.00

Home Link 1-11

4. Sample answer: Nadjia made one strip for 0 to 1 and one strip for 0 to 2. The strips she is comparing are not the same interval.

Home Link 1-12

1. Sample answer: Because $\frac{1}{3} = \frac{8}{24}$ and $\frac{1}{4} = \frac{6}{24}$, $\frac{7}{24}$ is between $\frac{1}{3}$ and $\frac{1}{4}$. A number line can always be broken into smaller parts.

Home Link 1-13

3. Greatest: $-4\frac{1}{6}$; Least: $-4\frac{5}{6}$

6a. 4   6b. $-2\frac{1}{2}$ 6c. $\frac{3}{4}$

Home Link 1-14

2. (1, 3)

Exploring Dot Plots and Landmarks

(1) Draw a dot plot for the following spelling test scores:

100, 100, 95, 90, 92, 93, 96, 90, 94, 90, 97

⟵————————————————————————————⟶

(2) The mode of the data in Problem 1 is _____.

(3) Draw a dot plot that represents data with the landmarks shown below.
Use at least 10 numbers.

Range: 7   Minimum: 6   Modes: 8 and 11

⟵————————————————————————————⟶

(4) Explain how you decided where to place your data on the dot plot in Problem 3.

_____

_____

_____

(5) Describe a situation the data in the dot plot in Problem 3 might represent.

_____

(6) Give the dot plot a title. Be sure to label the unit (for example, dollars or miles) for the number line.

Find an interesting graph on the Internet or in a newspaper or magazine.
Bring it to class tomorrow.

Using the Mean to Solve Problems

① Ms. Li brought pumpkin seed packs for her class. Each student received a pack. Her class predicted that there were 30 seeds in each pack. Here are the total number of seeds per pack the students in one group found when they counted: 20, 21, 23, 20, 22, 20.

Find the group mean for the pumpkin seed packs. _____ pumpkin seeds

② Another group in Ms. Li's class added their pumpkin seed counts to the data set. Here is what they have all together: 20, 21, 23, 20, 22, 20, 23, 27, 28, 29, 28, 27.

Make a dot plot for the combined data.

Pumpkin Seed Treats

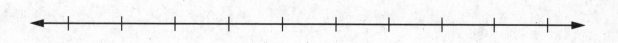

Find these landmarks for this data set.

Median: _____   Mode(s): _____   Mean: _____

③ If Ms. Li brought these packs every day for 20 days of class, about how many seeds would each student receive?

_____

Try This

④ If you were in charge of advertising these pumpkin seed packs, how many seeds would you advertise are in each pack? Why?

_____

Practice

⑤ 4 * 12 = _____   ⑥ _____ = 6 * 12

⑦ 15 * 5 = _____   ⑧ _____ = 13 * 4

7

Balancing Movies

Sandy asked six students how many movies they watched last month, and then graphed the results. The mean number of movies watched was 4.

① How likely is it that Sandy's graph would look like the graph at the right? Explain.

Number of Movies Watched

0 1 2 3 4 5 6 7 8

_____

_____

② Suppose that five students answered as shown. How many movies did the sixth student watch? _____

Plot the sixth point on the dot plot and explain how you know where to place it.

Number of Movies Watched

0 1 2 3 4 5 6 7 8

_____

③ Suppose that four students answered as shown. How many movies could the last two students watch?

_____

Plot the last two points on the dot plot and explain how you know where to place them.

Number of Movies Watched

0 1 2 3 4 5 6 7 8

_____

Try This

④ The data shown at the right is for four of six students surveyed. What two missing data points would make the mean 4?

Number of Movies Watched

_____

Plot the points on the dot plot.

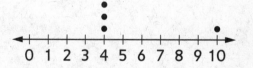

0 1 2 3 4 5 6 7 8 9 10

Practice   Solve.

⑤ $46 \div 2 =$ _____   ⑥ $80 \div 5 =$ _____   ⑦ $68 \div 2 =$ _____

Measures of Center

SRB
284-290

Math test scores (each out of 100 points) are shown below.

Mia's scores: 75, 75, 75, 85, 80, 95, 85, 90, 80, 80
Nico's scores: 55, 80, 90, 100, 70, 80, 50, 80, 75, 80

① Find the median and mean scores for each student.

Mia: Median _____   Mean _____

Nico: Median _____   Mean _____

② Which better represents each student's performance, the mean or median? Explain.

Mia: _____

_____

Nico: _____

_____

③ In their class, a score in the 80s is a B and a score in the 70s is a C. If their teacher uses the medians of their test scores to calculate grades, Mia and Nico would get the same grade. If the teacher uses the mean, Mia would get a B and Nico would get a C.

Explain how Mia's and Nico's scores have the same median and different means.

_____

_____

_____

④ If you were the teacher in Mia and Nico's class, would you use the median or the mean to calculate students' grades? Explain.

_____

_____

Practice   Solve.

⑤ 25 * 30 = _____   ⑥ _____ = 16 * 400

⑦ 150 * 600 = _____   ⑧ _____ = 90 * 130

11

Analyzing Persuasive Graphs

You are trying to convince your parents that you deserve an increase in your weekly allowance. You claim that during the past 10 weeks, the time you have spent doing jobs around the house (such as emptying the trash, mowing the lawn, and cleaning up after dinner) has increased. You have decided to present this information to your parents in the form of a graph. You have made two versions of the graph and need to decide which one to use.

SRB
308

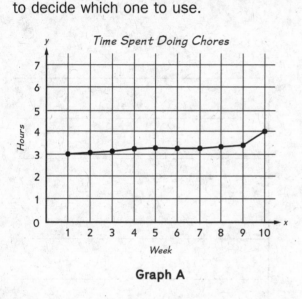

Graph A

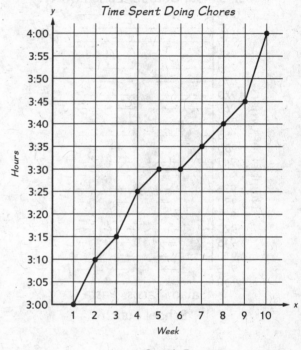

Graph B

(1) How are Graph A and Graph B similar?

_____

_____

(2) How are Graph A and Graph B different?

_____

_____

(3) Which graph, A or B, do you think will help you more as you try to convince your parents that you deserve a raise in your allowance? Why?

_____

_____

_____

13

Analyzing Persuasive Graphs (continued)

④ For the graph, describe what you plan to correct. Redraw the graph to give a more accurate picture of the data.

SRB
308

Correction(s): _____

Original Graph

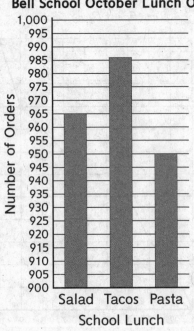

Bell School October Lunch Orders

My Corrected Version

Describe how your corrections changed what you see in the graph.

_____

_____

Practice

Solve.

⑤ $\frac{1}{12} + \frac{7}{12} =$ _____   ⑥ $\frac{3}{10} + \frac{7}{10} =$ _____   ⑦ $\frac{1}{8} + \frac{3}{8} =$ _____   ⑧ $1\frac{1}{2} + \frac{1}{2} =$ _____

Exploring Bar Graphs and Histograms

① Circle each graph that is a histogram.

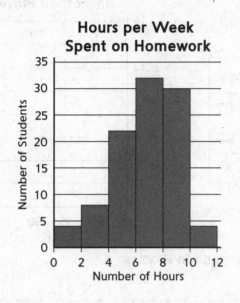

Hours per Week Spent on Homework

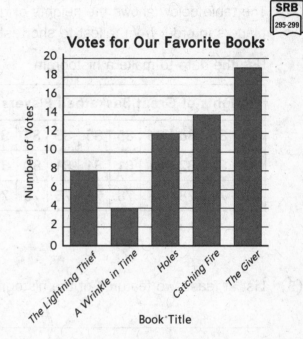

Votes for Our Favorite Books

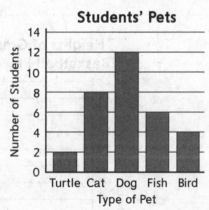

Students' Pets

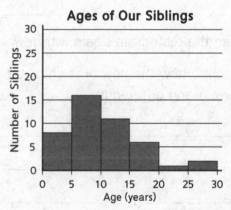

Ages of Our Siblings

② Pick one of the graphs above and list two different questions you could answer with the graph. Do not use the same kind of question twice (even about a different graph).

_____

_____

_____

③ Describe features of a graph that make it a histogram.

_____

_____

_____

15

Exploring Bar Graphs and Histograms (continued)

④ The table below shows the heights of great basketball players in order from tallest to shortest.

Use the data to make a histogram.

Heights of Great Basketball Players in Inches

89	88	86	86	85	85	85	84	84	84	82
82	82	82	81	81	81	81	80	80	80	80
79	79	78	78	78	75	75	73	70	70	69

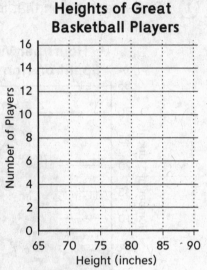

Heights of Great Basketball Players

⑤ List at least two features of the histogram you made in Problem 4.

_____

_____

⑥ Make the histogram again with fewer bins.

⑦ Describe how the new bins show the information in the graph differently.

_____

_____

_____

⑧ Why might you want to use one bin size instead of another to show data?

_____

_____

_____

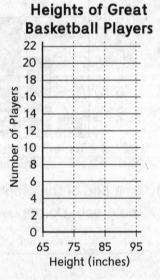

Heights of Great Basketball Players

Practice

Solve.

⑨ $\frac{7}{8} - \frac{3}{8} =$ _____

⑩ $\frac{9}{10} - \frac{6}{10} =$ _____

⑪ _____ $= \frac{4}{10} + \frac{6}{10}$

Kentucky Derby Winners

Use the graph of Kentucky Derby winners' times for the problems below.

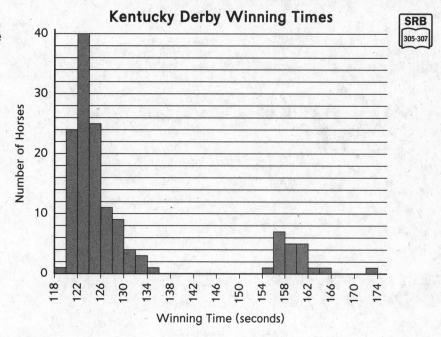

Kentucky Derby Winning Times

SRB
305-307

Number of Horses (y-axis)

Winning Time (seconds) (x-axis)

① Describe the shape of this graph.

_____

② Explain why the graph for this data set might have this shape.

_____

_____

③ Draw a line on the graph approximately where you think the mean is. Approximately where are the median and the mode compared to the mean?

_____

_____

Try This

④ Research and describe why the graph of Kentucky Derby winning times is this shape.

_____

Practice Solve.

⑤ _____ * 50 = 350   ⑥ 60 * 40 = _____   ⑦ 3,600 = 90 * _____

Exploring Histograms

Here are two histograms representing the lengths of the 20 longest rivers in the world.

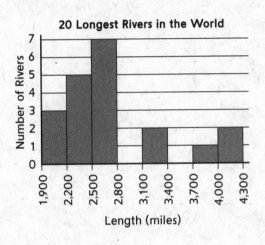

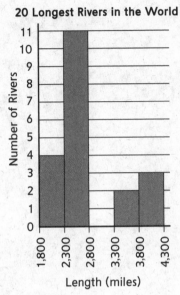

① Describe how the shapes of the graphs are different.

_____

② These histograms represent the same set of data. Why do they look different?

_____

③ **a.** Based on the graphs, what is the largest the range can be? _____

    b. Explain how you figured out the largest possible range.

    _____

    _____

④ **a.** Estimate the median for the lengths of the 20 longest rivers.

    _____

    b. Explain how you estimated the median.

    _____

    _____

19

Plotting Numbers

① Here is a list by month for the record low temperatures in Minneapolis, MN.
Plot the letters for the temperatures on the number line below.

SRB
94

A: January, −57°F   E: May, 4°F   I: September, 10°F

B: February, −60°F   F: June, 15°F   J: October, −16°F

C: March, −50°F   G: July, 24°F   K: November, −45°F

D: April, −22°F   H: August, 21°F   L: December, −57°F

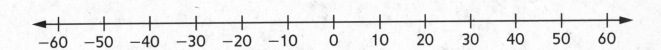

② A tree has a trunk, branches, and leaves above ground (positive)
and roots below ground (negative). Represent each height
as a point on the number line.

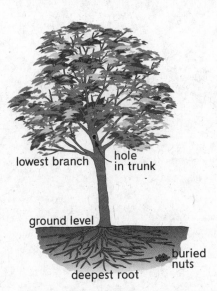

lowest branch   hole in trunk

ground level

deepest root   buried nuts

M: Lowest branch at 6 feet

N: Deepest root at 5 feet

P: Hole in trunk at 8 feet

Q: Ground level

R: Buried nuts at 3 feet

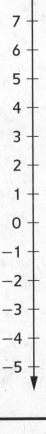

Practice

Solve.

③ $0.40 * 5 = _____   ④ $1.50 * 3 = _____

21

Fractions on a Number Line

(1) Find three rational numbers between each of the pairs of numbers below.

   a. $\frac{1}{3}$ and $\frac{5}{6}$ _____

         b. $\frac{1}{3}$ and $\frac{1}{5}$ _____

(2) **a.** Label the points on the number line.

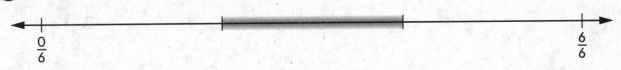

$\frac{0}{6}$ .... $\frac{6}{6}$

_____   _____

   b. Find two fractions in the highlighted section of the number line. _____

(3) **a.** Fill in the missing labels on the number line.

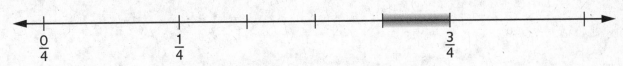

$\frac{0}{4}$ .... $\frac{1}{4}$ .... $\frac{3}{4}$ ....

_____   _____   _____   _____

   b. Find one fraction in the highlighted section of the number line. _____

(4) Nadjia created fraction strips to determine that $\frac{4}{5}$ is smaller than $\frac{3}{4}$.
Here is a sketch of her strips and how she lined them up. What mistake did she make?

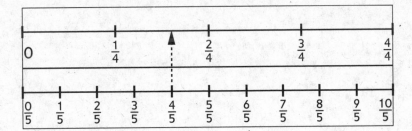

_____

_____

Practice

Solve.

(5) $\frac{3}{4} = \frac{\boxed{}}{16}$

(6) $\frac{18}{20} = \frac{\boxed{}}{10}$

(7) $\frac{\boxed{}}{7} = \frac{15}{21}$

Zooming In on the Number Line

(1) Maggie says there are no fractions between $\frac{1}{4}$ and $\frac{1}{3}$. Provide an example for Maggie and explain why you can always find another example.

_____

_____

_____

(2) One way to find fractions in between two fractions is to imagine zooming in on the number line. Insert the missing numbers for the number lines below.

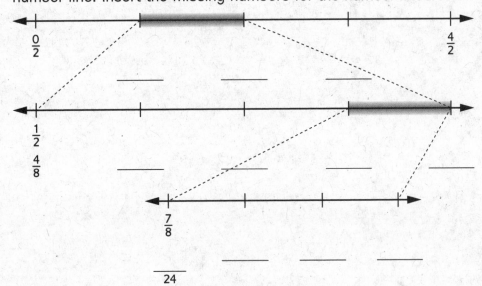

(3) Insert the missing numbers on the number line.

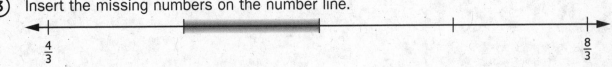

List at least three fractions that are in the highlighted section of the number line.

_____

Practice

Write an equivalent fraction.

(4) $4\frac{1}{3}$ _____

(5) $5\frac{2}{9}$ _____

(6) $2\frac{5}{6}$ _____

Negative Numbers on a Number Line

① Plot the following points.

$A: -1\frac{3}{4}$ $B: 4\frac{1}{6}$ $C: 0$ $D: \frac{4}{5}$ $E: -3$ $F: 3\frac{2}{3}$ $G: -\frac{2}{3}$

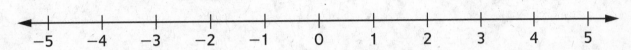

② **a.** On the vertical number line, label the topmost and bottommost tick marks as -4 and -5. Label the topmost tick mark with the greater value.

     b. Plot and label the following points as accurately as you can. $-4\frac{1}{2}$, $-4\frac{1}{3}$, $-4\frac{2}{3}$, $-4\frac{1}{4}$, $-4\frac{2}{4}$, $-4\frac{3}{4}$, $-4\frac{1}{6}$, $-4\frac{2}{6}$, $-4\frac{3}{6}$, $-4\frac{4}{6}$, $-4\frac{5}{6}$

③ Of the points you plotted on the number line in Problem 2b, which has the greatest value?_____

    Which has the least value? _____

④ How can you use a number line to compare values?

_____

_____

For Problems 5–6, you may draw a number line to help you.

⑤ Write two numbers that fit each description.

     a. Between -1 and -2 _____

     b. Less than -3 _____

⑥ Write the opposite of each number.

     a. -4 _____   **b.** $2\frac{1}{2}$ _____   **c.** $-\frac{3}{4}$ _____

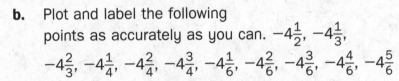

Practice

Write the first three multiples of each number.

⑦ 9 _____   ⑧ 7 _____   ⑨ 21 _____

Plotting Points on a Coordinate Grid

SRB
95-96

① Plot the following points on the coordinate grid. Label each with its letter.

A: School: (8, 0)   B: Library: (8, 5)   C: Park: (1, −3)

D: Grocery store: (−4, −2)   E: My house: (6, 3)

F: Post office: (1, 9)

G: Bank: (−5, 7)

H: Friend's house: (−4, 3)

② You walk a straight line from your house to your friend's house. Plot the point that is halfway between the two houses. Label this point M.

Write the ordered pair for point M. _____

③ Plot and label two points on the coordinate grid. Place your points in different quadrants.

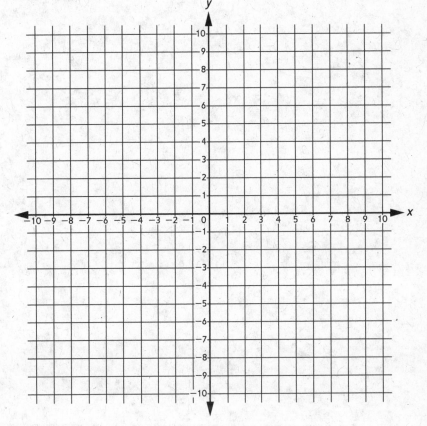

Letter: _____   Location: _____   Ordered pair: _____

Letter: _____   Location: _____   Ordered pair: _____

④ Explain how to plot the point (−3, 5).

_____

_____

Practice

List all of the factors.

⑤ 14 _____   ⑥ 20 _____

⑦ 17 _____   ⑧ 32 _____

Fraction Operations and Ratios

Numbers, in all their forms, are important in our everyday lives. Unit 2 focuses on the uses of fractions and ratios. The unit begins with a review of factors and multiples. Students learn to find the greatest common factors and least common multiples for pairs of numbers, and they use these concepts to solve real-world and mathematical problems.

Students revisit and apply fraction multiplication. Many of the fractions your child will work with (halves, thirds, fourths, sixths, eighths, tenths, twelfths, and sixteenths) are common in everyday situations. Students explore properties and identify patterns in fraction multiplication—for example, they find that when one factor is a fraction less than 1, the product is less than one or both factors. Students also review fraction-multiplication models and algorithms they used in *Fifth Grade Everyday Mathematics,* such as the area model shown at right.

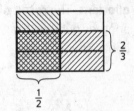

Students review fraction division in the context of real-world and mathematical problems. They are encouraged to use pictures and to estimate in order to make sense of the problems, and they use a common-denominator algorithm to divide fractions. The emphasis of the fraction-division lessons is on understanding the concept of fraction division and investigating when and how to use fraction division. For example, students solve problems such as this: For how many days will $2\frac{1}{2}$ gallons of milk last if a family consumes $\frac{1}{2}$ gallon each day? Finally, students work with reciprocals and solve fraction-division problems using the U.S. traditional algorithm (multiplying the dividend of the division problem by the reciprocal of the divisor).

The second half of Unit 2 is devoted to exploring ratios. Ratios use division to compare quantities. Rates, which are ratios that compare quantities with different units, constitute part of this exploration. In this unit, your child uses ratio representations and ratio notation in many forms to make a variety of comparisons. For example, if 5 eggs out of a dozen are brown and the rest are white, then the ratio of brown eggs to white eggs is 5 : 7 (or 5 to 7), and the ratio of brown eggs to the total number of eggs is 5 : 12 (or 5 out of 12, or $\frac{5}{12}$). Students make and use ratio/rate tables to solve real-world and mathematical problems, including measurement-conversion problems.

Please keep this Family Letter for reference as your child works through Unit 2.

Vocabulary

Important terms in Unit 2:

dividend The number in division that is being divided. For example, in $35 \div 5 = 7$, the dividend is 35.

divisor In division, the number that divides another number. For example, in $35 \div 5 = 7$, the divisor is 5.

greatest common factor (GCF) The largest factor that two or more counting numbers have in common. For example, the common factors of 24 and 36 are 1, 2, 3, 4, 6, and 12, and their greatest common factor is 12.

least common multiple (LCM) The smallest number that is a multiple of two or more given numbers. For example, common multiples of 6 and 8 include 24, 48, and 72. The least common multiple of 6 and 8 is 24.

prime number A counting number that has exactly two different factors, itself and 1.

quotient The result of dividing one number by another number. For example, in $35 \div 5 = 7$, the quotient is 7.

ratio A comparison of two quantities using division. Ratios can be stated in words or expressed as fractions, decimals, or percents. Ratios can also be written as two numbers separated by a colon. For example, if a team wins 3 games out of 5 games played, the ratio of wins to total games is $\frac{3}{5}$, 3/5, 0.6, 60%, 3 to 5, or 3 : 5 (read "three to five").

reciprocals Two numbers whose product is 1. These numbers are also called *multiplicative inverses*. For example, the reciprocal, or multiplicative inverse, of 5 is $\frac{1}{5}$.

Do-Anytime Activities

1. Help your child look for examples of fractions in advertisements, brochures, newspapers, magazines, recipes, and other sources. Discuss how the examples you found use fractions.

2. Help your child practice using ratios by comparing sports teams' records. For example, compare wins to losses (such as 15 to 5) or wins to games played (15 out of 20).

3. Look for enlarged or reduced images of objects. For example, a cereal box may display a picture of the cereal that is two times its actual size, or a science magazine may show pictures of insects or bacteria that have been enlarged several hundred times their actual size. Have your child explain to you what the size-change factor means.

Building Skills through Games

In this unit, your child will work on a variety of skills by playing the following games:

Fraction Top-It (**Multiplication**) Students find and compare fraction products.

Divisibility Dash Students recognize multiples and use divisibility tests.

Division Top-It Students solve division problems and compare quotients.

Fraction Capture Students find and rename equivalent fractions.

High-Number Toss (**Decimal**) Students compare and write decimals.

Multiplication Bull's Eye Students estimate products of 2- and 3-digit numbers.

Landmark Shark Students find mean, median, and mode(s).

Hidden Treasure Students plot and identify points on a coordinate grid.

As You Help Your Child with Homework

As your child brings assignments home, you may want to go over the instructions together, clarifying them as necessary. The answers listed below will guide you through the unit's Home Links.

Home Link 2-1

1. **a.** 14 **b.** 16 **c.** 8 **d.** 5

3. $\frac{48 \div 16}{64 \div 16} = \frac{3}{4}$

4. **a.** 7 bracelets
   b. 4 blue beads and 3 red beads

6. 6

Home Link 2-2

1. **a.** 30 **b.** 60 **c.** 30 **d.** 35

2. **a.** 25; 300 **b.** 12; 144

6. **a.** 10:30 P.M.
   b. Sample answer: I started by making lists of multiples, and then I realized I was finding the LCM.

8. 16,800 9. 3,500 10. 9,600

Home Link 2-3

1. Sample answers: $\frac{6}{12}$; $\frac{1}{2}$

2. $\frac{1}{2} * \frac{2}{3} = \frac{2}{6}$ 4. $\frac{3}{4} * \frac{1}{4} = \frac{3}{16}$

5. 30 6. 10 7. 6

Home Link 2-4

1. $\frac{1}{2}$ of the garden; $\frac{2}{3} * \frac{3}{4} = \frac{6}{12}$ (or $\frac{1}{2}$)

2. $\frac{3}{5}$ of the garden; $\frac{1}{5} * 3 = \frac{3}{5}$

3. Sample estimate: 15 lb; $5\frac{1}{3} * 2\frac{2}{3} = 14\frac{2}{9}$

4. 9 5. 10 6. 24 7. 2

Home Link 2-5

1. **a.** $\frac{5}{4}$ **b.** 1 **c.** 10 **d.** 10 **e.** 6 **f.** 2

3. $\frac{2}{3} * 6 = 4$ pages   4. $8\frac{1}{2} * 10\frac{2}{3} = 90\frac{2}{3}$ ft²

5. 6,613 6. 8,448 7. 10,872

Home Link 2-6

1. 6 batches

2. **a.** $\frac{6}{8} \div \frac{3}{8} = 2$ **c.** $\frac{36}{8} \div \frac{4}{8} = 9$

3. $15 \div 1\frac{1}{2} = 10$; 10 necklaces

4. $4\frac{1}{2} \div \frac{3}{4} = 6$; 6 meters

5. 10 6. 2 7. 18

Home Link 2-7

1. Less than $4\frac{2}{3}$ feet; $4\frac{2}{3} \div 7$; $\frac{2}{3}$ foot

2. Less than $5\frac{1}{4}$ feet; $10\frac{1}{2} \div 5\frac{1}{4}$; 2 feet

3. More than 5 dogs; $5\frac{1}{4} \div \frac{3}{4}$; 7 dogs

4. 21 5. 8 6. 20

Home Link 2-8

1. $\frac{3}{1} * \frac{3}{2} = \frac{9}{2}$ 2. $\frac{1}{5} * \frac{9}{8} = \frac{9}{40}$

3. $\frac{4}{1} * \frac{7}{5} = \frac{28}{5}$ 4. $\frac{5}{3} * \frac{5}{3} = \frac{25}{9}$

5. $\frac{2}{5} * \frac{4}{3} = \frac{8}{15}$ 6. $\frac{3}{5} * \frac{1}{4} = \frac{3}{20}$

7. $3 \div \frac{1}{4}$; $3 * 4$; 12 servings

9. $8\frac{2}{3} \div \frac{1}{3}$; $\frac{26}{3} * 3$; 26 name tags

10. $7.50 11. $1.80 12. $8.10

Home Link 2-9

1. 7 female : 9 puppies; 2 male : 7 female

2. 6; Sample answer: 6 white tiles : 15 tiles

3. 18; Sample answer: 18 white : 6 shaded

4. Sample answer: 3 white : 1 shaded; 9

5. 7 Special Swirls

6. Sample answer: 4 : 7; 140   7. $\frac{15}{24}$ 8. 1   9. $\frac{16}{63}$

Home Link 2-10

1. **a.** 12 green tiles   **b.** 40 white tiles
   c. 10 green tiles
   d. 14 green tiles and 35 white tiles

2. Sample answer: 3 to 7 requires either 60 or 70 tiles.

3. 4   4. $\frac{1}{4}$ 5. 14

Home Link 2-11

2. 10 mm : 1 cm; 50 mm : 5 cm;
   3,000 mm : 300 cm; 250 mm : 25 cm

4. **b.** Sample answer: Multiply the dimensions of the 2-by-3 rectangle by 2 to get 4 and 6. The ratios of width to length are equivalent.

   c. $\frac{2}{4}, \frac{2}{3}, \frac{4}{6}, \frac{1}{5}$

5. $11.92   6. $74.85

Home Link 2-12

1. **a.** Recipe B
   c. Sample answer: Recipe B has 2 groups of raspberries and watermelon in the same ratio as Recipe A but with extra raspberries.

2. Sample answer: 10 cups raspberries, 12 cups watermelon

3. Sample answer: 1 cup raspberries, 3 cups watermelon

4. $\frac{2}{5}$ cup raspberries, $\frac{3}{5}$ cup watermelon

5. 80   6. 421   7. 310

Home Link 2-13

1. Sample answer: Miles per hour, cost per item, words typed per minute

2. 12 cans of water   3. $13\frac{1}{3}$ mi

4. 24 min   5. $\frac{1}{3}$ cup of lime juice

6. >   7. >   8. >

Home Link 2-14

1. **a.** 20; 5; 15   **b.** 4; 16; 24   **c.** 3; 9; 21

3. Sample answer: Jada's snail is the fastest. The line gets higher faster.

4. >   5. >   6. <

Finding the Greatest Common Factor

SRB
105

①   Use any method to find the greatest common factor for the number pairs.

    a.   GCF (42, 56) = _____   **b.**   GCF (32, 80) = _____

    c.   GCF (72, 16) = _____   **d.**   GCF (10, 40, 25) = _____

②   Explain how you found GCF (42, 56) in Problem 1a.

_____

_____

③   Use the GCF to find an equivalent fraction for $\frac{48}{64}$. Show your work.

    Answer: _____

④   Jenny will use 28 blue beads and 21 red beads to make identical bracelets.

    a.   What is the greatest number of bracelets she can make?

       _____

    b.   How many blue beads and how many red beads will be on each bracelet?

       _____

⑤   Explain how a set of numbers can have a GCF greater than 1.

_____

_____

Try This

⑥   GCF (12, 24, 30, 42) = _____

Practice

Insert the missing digits to make each number sentence true.

⑦   ___,___63 − 3,9___9 = 2,83___   ⑧   71,___4___ − 4,8___6 = 6___,270

Least Common Multiple

1 Find the least common multiple for each pair of numbers.

   a. LCM (10, 15) = _____   **b.** LCM (12, 15) = _____

   c. LCM (6, 10) = _____   **d.** LCM (7, 5) = _____

2 Find the greatest common factor and least common multiple for each pair of numbers.

   a. GCF (75, 100) = _____   **b.** GCF (36, 48) = _____

      LCM (75, 100) = _____   LCM (36, 48) = _____

Use the LCM to find equivalent fractions with the least common denominator.

3 $\frac{3}{4}$ and $\frac{5}{6}$ **4** $\frac{1}{6}$ and $\frac{3}{8}$ **5** $\frac{4}{25}$ and $\frac{4}{15}$

LCM (4, 6) = _____   LCM (6, 8) = _____   LCM (25, 15) = _____

Fractions: _____   Fractions: _____   Fractions: _____

6 **a.** On a website, there is an ad for jeans every 5 minutes, an ad for sneakers every 10 minutes, and an ad for scarves every 45 minutes.
If they all appeared together at 9:00 P.M., when is
the next time they will all appear together? _____

   b. Explain how you used GCF or LCM to solve the problem.

_____

_____

7 Explain why the LCM is at least as large as the GCF.

_____

_____

Practice

Estimate.

8 5,692 * 3 = _____   **9** 69 * 54 = _____   **10** 78 * 123 = _____

Fraction-Multiplication Review

Represent the problem on a number line, and then solve the problem.

① $\frac{2}{3} * \frac{9}{12} =$ _____

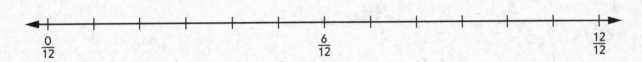

$\frac{0}{12}$ $\frac{6}{12}$ $\frac{12}{12}$

② Maliah has $\frac{2}{3}$ cup of raisins. She used $\frac{1}{2}$ of her raisins to make muffins. What fraction of a cup of raisins did she use?

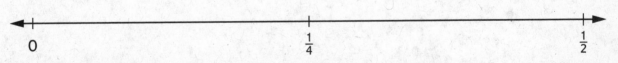

Number sentence: _____

③ On the back of this page, write and solve a number story for $\frac{1}{4} * \frac{1}{2}$.

Try This

④ Ryse sprinted $\frac{3}{4}$ of a lap around the running track at school. A whole lap is $\frac{1}{4}$ mile.

How far did he sprint?

0                              $\frac{1}{4}$ $\frac{1}{2}$

Number sentence: _____

Practice

Estimate.

⑤ 845 ÷ 24 = _____   ⑥ 6,450 ÷ 639 = _____   ⑦ 129 ÷ 19 = _____

39

Companion Gardening

Draw and label area models and write number sentences to represent and solve Problems 1–2.

SRB
188-192

(1) In companion planting, marigold flowers are used to repel insects that harm melon plants. Community gardeners plant $\frac{2}{3}$ of a rectangular garden bed with melon plants. They plant $\frac{3}{4}$ of the melon area with marigolds.

What fraction of the garden bed will have both plants growing together? _____

Number sentence: _____

(2) Two plants that grow well together are tomatoes and basil. This year, $\frac{1}{5}$ of a garden bed was planted with tomatoes and basil. Next year, the area will be 3 times as large.

What will the area be next year? _____

Number sentence: _____

First estimate, then use a partial-products diagram to solve Problem 3.

(3) Last year a community garden produced $5\frac{1}{3}$ pounds of carrots. This year, better weather resulted in a harvest $2\frac{2}{3}$ times as large. How many pounds of carrots were harvested this year?

Estimate: _____

Number sentence: _____

Practice Find equivalent fractions.

(4) $\frac{3}{4} = \frac{\boxed{}}{12}$ **(5)** $\frac{18}{20} = \frac{9}{\boxed{}}$ **(6)** $\frac{6}{7} = \frac{\boxed{}}{28}$ **(7)** $\frac{24}{36} = \frac{\boxed{}}{3}$

Fraction Multiplication

Mara's strategy: $\frac{6}{8} * \frac{2}{3} = \left(6 * \frac{1}{8}\right) * \left(2 * \frac{1}{3}\right)$

$= (6 * 2) * \left(\frac{1}{8} * \frac{1}{3}\right)$

$= 12 * \frac{1}{24}$

$= \frac{12}{24}$

SRB
188-192

1. Use Mara's strategy to rename the fractions as whole numbers and unit fractions. Then group your factors to make the problem easier. Show the steps you use.

   a. $\frac{5}{2} * \frac{2}{4} =$ _____

   b. $\frac{10}{8} * \frac{8}{10} =$ _____

   c. _____ $= 12 * \frac{5}{6}$

   d. _____ $= \frac{5}{2} * 4$

   e. $\frac{21}{3} * \frac{6}{7} =$ _____

   f. $9 * \frac{2}{9} =$ _____

2. Choose two problems from above that are alike in some way. Describe how they are alike.

   _____

   _____

Use any model or strategy to solve Problems 3–4. Write a number sentence.

3. Samantha had 6 pages of homework. She finished $\frac{2}{3}$ of her assignment.

   How many pages did she finish?

   Number sentence:

   _____

4. A room measures $8\frac{1}{2}$ feet by $10\frac{2}{3}$ feet.

   What is the area of the room?

   Number sentence:

   _____

Practice

5. $389 * 17 =$ _____

6. _____ $= 176 * 48$

7. $453 * 24 =$ _____

Division Using
Common Denominators

(1) Draw a picture or diagram and solve the problem.

Rudi has 4 cups of almonds.
His trail mix recipe calls for $\frac{2}{3}$ cup of almonds.
How many batches of trail mix can he make?

_____

(2) Use common denominators to solve the problems.

Write a number sentence to show how you rewrote the problem with common denominators.

Check your answers.

a. $\frac{3}{4} \div \frac{3}{8} =$ _____   Number sentence: _____

b. $3\frac{1}{3} \div \frac{5}{6} =$ _____   Number sentence: _____

c. $\frac{36}{8} \div \frac{1}{2} =$ _____   Number sentence: _____

(3) Michelle is cutting string to make necklaces.
She has 15 feet of string. She needs $1\frac{1}{2}$ feet of string for each necklace.
How many necklaces can she make?

Number model: _____   Solution: _____

(4) A rectangular window has an area of $4\frac{1}{2}$ square meters. Its width is $\frac{3}{4}$ meter.
What is its length?

Number model: _____   Solution: _____

Practice

Solve.

(5) GCF (20, 30) = _____   **(6)** GCF (6, 16) = _____   **(7)** GCF (36, 54) = _____

More Exploring Fraction Division

For problems 1–3, circle the best estimate and the correct number model.
Then solve the problem.

SRB
193-196

(1) Stan is in woodworking class with 6 friends.
They have to split a board that is $4\frac{2}{3}$ feet long equally among the seven of them.
How long will each person's piece be?

Estimate: More than $4\frac{2}{3}$ feet   Less than $4\frac{2}{3}$ feet

Number model: $7 \div 4\frac{2}{3}$ $4\frac{2}{3} \div 7$

Answer: _____

(2) The area of a rectangle is $10\frac{1}{2}$ square feet. The length is $5\frac{1}{4}$ feet.
How wide is the rectangle?

Estimate: More than $5\frac{1}{4}$ feet   Less than $5\frac{1}{4}$ feet

Number model: $10\frac{1}{2} \div 5\frac{1}{4}$ $5\frac{1}{4} \div 10\frac{1}{2}$

Answer: _____

(3) Sounya walks dogs on Saturdays. It takes $\frac{3}{4}$ of an hour to walk each dog.
She has $5\frac{1}{4}$ hours. How many dogs can she walk?

Estimate: More than 5 dogs   Fewer than 5 dogs

Number model: $\frac{3}{4} \div 5\frac{1}{4}$ $5\frac{1}{4} \div \frac{3}{4}$

Answer: _____

Practice

Find the LCM.

(4) LCM (3, 7) = _____   (5) LCM (8, 4) = _____   (6) LCM (10, 4) = _____

Fraction Division

Rewrite and solve the division problems using the Division of Fractions Property.

SRB
196

Example: $\frac{3}{8} \div \frac{2}{5} = \frac{3}{8} * \frac{5}{2} = \frac{15}{16}$

① $3 \div \frac{2}{3} =$ _____

② $\frac{1}{5} \div \frac{8}{9} =$ _____

③ $4 \div \frac{5}{7} =$ _____

④ $1\frac{2}{3} \div \frac{3}{5} =$ _____

⑤ $\frac{2}{5} \div \frac{3}{4} =$ _____

⑥ $\frac{3}{5} \div 4 =$ _____

⑦ How many $\frac{1}{4}$-cup servings of cottage cheese are in a 3-cup container?

Division number model: _____   Multiplication number model: _____

Solution: _____

⑧ Philip went on a $3\frac{1}{2}$-mile hike. He hiked for 2 hours.
About how far did he go in 1 hour?

Division number model: _____   Multiplication number model: _____

Solution: _____

⑨ Adam is using ribbon to decorate name tags for the class picnic.
He has $8\frac{2}{3}$ feet of blue ribbon. He needs $\frac{1}{3}$ foot of ribbon for each name tag.
How many name tags can he decorate?

Division number model: _____   Multiplication number model: _____

Solution: _____

Practice

Add or subtract.

⑩ $\$4.50 + \$3 =$ _____

⑪ $\$5.00 - \$3.20 =$ _____

⑫ _____ $= \$6.30 + \$0.45 + \$1.35$

Using Ratios to Represent Situations

① Lenore's dog gave birth to a litter of 9 puppies.
Two of the puppies are male. Write ratios for the following:
Number of female puppies to the total number of puppies _____

Number of male puppies to female puppies _____

For Problems 2–4, draw a picture to help you solve the problem. Record a ratio.

② There are 15 tiles. 2 out of 5 tiles
are white. How many tiles are white? _____
Write the ratio of white tiles to total tiles.

_____

③ There are 24 tiles. 3 out of 4 tiles
are white. How many tiles are white? _____
Write the ratio of white tiles to shaded tiles.

_____

④ There are 3 times as many white tiles as
there are shaded tiles. Write this ratio.

_____

How many tiles are white if there are 12 tiles in total? _____

⑤ The Mighty Marble Company fills bags of marbles
with a ratio of 3 Special Swirls out of every 9 marbles.
How many Special Swirls are in a bag that has 21 marbles? _____

Try This

⑥ One class of 28 students has a ratio of 3 girls to 4 boys. What is the ratio for the
number of boys to total number of students in the class?

_____

There are 60 girls in the whole sixth grade and the ratio is the same. How many

students are there in sixth grade? _____

Practice Solve.

⑦ $\frac{5}{6} * \frac{3}{4} =$ _____

⑧ $\frac{2}{3} * 1\frac{1}{2} =$ _____

⑨ $\frac{8}{9} * \frac{2}{7} =$ _____

More with Tape Diagrams

Draw tape diagrams to solve the problems. Label your diagrams and your answers.

SRB
45-48

① Frances is helping her father tile their bathroom floor. They have tiles in two colors: green and white. They want a ratio of 2 green tiles to 5 white tiles.

   a. They use 30 white tiles.
How many green tiles do they use?

     _____

   b. How many white tiles would they need if they use 16 green tiles?

     _____

   c. They use 35 tiles in all.
How many are green?

     _____

   d. They use 49 tiles. How many of each color did they use?

     _____

   e. Explain how you used the tape diagram to solve Part d.

     _____

     _____

Try This

② Frances and her father decide to also tile their kitchen floor. For every 3 white tiles they plan to use 7 green tiles. The kitchen floor has room for 63 tiles total. Explain why they cannot cover the kitchen floor using the ratio 3 : 7.

   _____

   _____

Practice Divide.

③ $\frac{4}{5} \div \frac{1}{5} =$ _____

④ $\frac{1}{5} \div \frac{4}{5} =$ _____

⑤ $7 \div \frac{1}{2} =$ _____

Finding Equivalent Ratios

Use the pictures to help you figure out the equivalent ratios.

SRB
41

①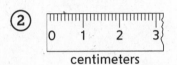

1 foot

Ratio of feet to inches:

1 foot : _____ inches

3 feet : _____ inches   7 feet : _____ inches   _____ feet : 144 inches

②

centimeters

Ratio of millimeters to centimeters:

10 mm : _____ cm

_____ mm : 5 cm   _____ mm : 300 cm   250 mm : _____ cm

③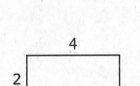

Ratio of legs to spiders:

8 legs : _____ spider

_____ legs : 4 spiders   _____ legs : 9 spiders   320 legs : _____ spiders

④ **a.** Circle the similar rectangles.

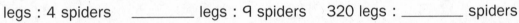
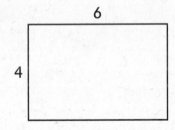
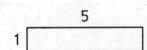

6

4   3   4   5

2   2   1

b. Explain why the rectangles you circled are similar.

_____

_____

c. Under each rectangle, use fraction notation to write the width-to-height ratio.

Practice Multiply mentally to find the cost.

⑤ 4 pens at $2.98 each _____   ⑥ 3 books at $24.95 each _____

55

Using Ratios to Make Fruit Cups

Oliver has two fruit-cup recipes that have different ratios of raspberries and watermelon.

Recipe A	Recipe B
2 cups raspberries	5 cups raspberries
3 cups watermelon	11 cups total

① **a.** Which fruit-cup recipe would have a stronger raspberry taste? _____

b. Draw a picture or diagram to support your answer.

c. Explain how your picture or diagram supports your answer.

_____

_____

② Create a fruit-cup recipe that would taste the same as Recipe B, but uses more than 11 cups of fruit.

List your ingredients: _____

③ Create a fruit-cup recipe that would make a fruit cup with a weaker raspberry taste than Recipes A and B.

List your ingredients: _____

Try This

④ If you only want 1 cup of fruit salad made from Recipe A, what measurements of watermelon and raspberries do you need?

_____

Practice Divide.

⑤ 560 ÷ 7 = _____   ⑥ 842 ÷ 2 = _____   ⑦ 930 ÷ 3 = _____

Ratio/Rate Tables and Unit Rates

① List three examples of a rate:

_____

Draw a ratio/rate table to solve each problem. The first table has been drawn for you, but it is not complete.

② One 12-ounce can of frozen juice is mixed with three 12-ounce cans of water. How many cans of water do you need for 4 cans of juice?

Cans of Water	3	
Cans of Juice	1	4

_____

③ A hiker's map has a scale of 3 inches to 10 miles.
The trail is 4 inches long on the map. How long is the actual hike? _____

④ Amy types 125 words in 2 minutes.
About how long will it take her to type a 1,500-word report? _____

Try This

⑤ A recipe for lime salad dressing calls for $\frac{1}{4}$ cup lime juice and $\frac{3}{4}$ cup olive oil.

How much lime juice would you use with 1 cup olive oil? _____

Practice Record >, <, or =.

⑥ −3 _____ −5   ⑦ 6 _____ −7   ⑧ −8 _____ −9

Graphing Rates

SRB
51-52

Snails move slowly. Jada, Reality, and Barb had a snail race.
Then they compared the rates at which the snails crawled.

① Fill in the ratio/rate table with equivalent rates.

a. Jada's Snail

Minutes	10			
Inches	4	8	2	6

b. Reality's Snail

Minutes	12			
Inches	3	1	4	6

c. Barb's Snail

Minutes	15			
Inches	5	1	3	7

② Treat each rate as an ordered pair. Graph each snail's rate using a different color.

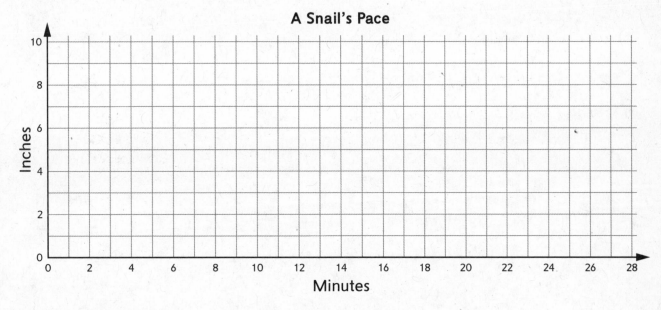

A Snail's Pace

③ Which snail is the fastest? Use the graph to explain how you know.

_____

_____

Practice Insert >, <, or = to make each sentence true.

④ 7 ____ 4.65   ⑤ 0.1 ____ 0.01   ⑥ 0.205 ____ 0.22

61

Decimal Operations and Percent

The first part of Unit 3 is devoted to the continued study of decimal operations from earlier grades. Your child will use estimation strategies, mental-math methods, and paper-and-pencil algorithms to work with decimal numbers. Students will connect their understanding of fractions from Unit 2 to finding fraction–decimal equivalencies. They will review comparing and ordering decimals, as well as writing decimals using expanded notation. *For example:*

$$3.214 = 3 * (1) + 2 * \left(\frac{1}{10}\right) + 1 * \left(\frac{1}{100}\right) + 4 * \left(\frac{1}{1,000}\right)$$

$$= 3 * (1) + 2 * (0.1) + 1 * (0.01) + 4 * (0.001)$$

Computation is an important part of problem solving. Students review procedures for decimal addition, subtraction, and multiplication, and this unit introduces division of decimals. Your child will apply computation procedures in a variety of real-world and problem-solving situations. Students also interpret decimal remainders and determine when it is appropriate to ignore them and when it is not. *For example:*

- Ron has $19.60. He wants to buy as many notebooks as possible. Each notebook costs $3.50. How many notebooks can he buy?

  $19.60 \div 3.50 = 5.6$

  Ignore the decimal remainder, because you cannot buy fractional notebooks.

  Ron can buy 5 notebooks.

- Marcy has lunch with 5 friends. The bill is $51.00. How much will each person pay?

  $51.00 \div 6 = 8.5$

  Use the decimal remainder, because you can have fractional amounts of money.

  Each person will pay $8.50.

This unit focuses on building facility with fraction–decimal–percent equivalents. Percents are introduced as a special kind of ratio: *percent* means "per 100." A percent is the ratio of some number out of 100. Students investigate the concept of percent, and they explore and solve real-world problems involving percents.

In Unit 3, students apply their work with percents to analyze and create box plots. (*See the Vocabulary section of this letter for an example.*) The "box" in a box plot represents the middle 50% of the data values. The line segment (sometimes called a "whisker") extending to the left represents the bottom 25% of the data values, and the line segment extending to the right represents the top 25% of the data values.

Please keep this Family Letter for reference as your child works through Unit 3.

Vocabulary

Important terms in Unit 3:

box plot (Also sometimes called a box-and-whiskers plot) A plot displaying the spread, or distribution, of a data set using 5 landmarks: minimum, lower quartile, median, upper quartile, and maximum. (*See below for explanations of quartiles.*) Together with the median, the quartiles split the data into four quarters.

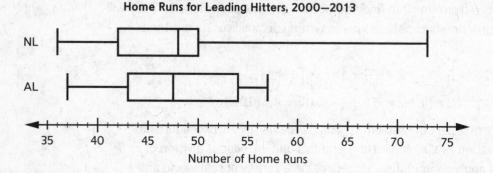

lower quartile (Q1) The middle value of the bottom half of a data set. This is shown on the box plot as the lower boundary of the box. In the example above, the lower quartile for the National League is at 42, and for the American League it is at 43. Lower quartile can also refer to the group of values between the minimum and the single Q1 value.

percent (%) Per hundred, for every hundred, or out of a hundred. For example, *48% of the students in the school are boys* means that out of every 100 students in the school, 48 are boys. Also, $1\% = \frac{1}{100} = 0.01$.

upper quartile (Q3) The middle value of the top half of the data set. This is shown on the box plot as the upper boundary of the box. In the example above, the upper quartile for the National League is at 50, and for the American League it is at 54. Upper quartile can also refer to the group of values between the maximum and the single Q3 value.

Do-Anytime Activities

Work with your child on some of the concepts taught in Unit 3:

1. As your child is learning about decimals and percent in the classroom, help your child find examples in the everyday world—newspaper advertisements, finance or sports sections of a newspaper, statistical reports, food-calorie guidelines, interest rates, and so on.

2. Practice using percents in the context of tips. Invite your child to find the tip the next time your family goes to a restaurant. For example, have your child calculate $\frac{1}{10}$, or 10%, of the bill and use this to find 5% of the bill. Then combine these amounts to find a 15% tip.

3. Use a loan statements to help your child understand the importance of interest rates and how they affect your monthly payments. Your child can also practice decimal operations by calculating the accuracy of utility or cell phone bills.

4. Encourage your child to read nutrition labels and calculate the percent of calories from fat in the item.

$$\text{fat calories / total calories} = \frac{?}{100} = ?\% \text{ of calories from fat}$$

If your child enjoys this activity, extend it by figuring the percent of calories from other nutrients.

Building Skills through Games

In this unit, your child will work on a variety of skills by playing the following games:

Getting to One
Students estimate quotients and compare decimals.

Multiplication Wrestling (Mixed Numbers)
Students multiply mixed numbers.

Fraction Action, Fraction Friction
Students estimate sums and add fractions.

Hidden Treasure
Students plot and identify points on a coordinate grid.

Division Top-It
Students estimate quotients.

Name That Number
Students name numbers with expressions and practice computation skills.

Ratio Dominoes
Students find equivalent ratios.

As You Help Your Child with Homework

As your child brings home assignments, you might want to go over the instructions together, clarifying them as necessary. The answers listed below will guide you through the unit's Home Links.

Home Link 3-1

1. Sample answers: **a.** 0.20, 0.200 **b.** 0.130, 0.130000 **c.** 2.1450, 2.1450000 **d.** 7.060, 7.06000

2. **a.** = **b.** < **c.** < **4.** 0.21, 0.210

5. **a.** Cross out $\left(2 * \frac{1}{100}\right) + \left(3 * \frac{1}{1,000}\right)$ and $(2 * 0.01) + (3 * 0.001)$. Sample answers: $\frac{20}{100} + \frac{3}{100}$; $23 * \frac{1}{100}$

   b. Cross out $\frac{10,045}{10,000}$; $10,045 * 0.01$; and $10,045 * 1$. Sample answers: $10,045 * 1$; $0.010045 * 1,000$

Home Link 3-2

2. Sample answers given. Hop Set 1: $(2 * 0.1) + (8 * 0.01) = 0.28$; Hop Set 2: $28 * 0.01 = 0.28$

3. **a.** $(3 * 1) + (4 * 0.1) + (8 * 0.01)$

   b. Sample answer: 3.486

4. 8.0323, 8.0329, 8.032222

Home Link 3-3

1. 3.3 min **2.** 5.82 m **3.** 754.41 **4.** 176.263 mph

Home Link 3-4

1. 89.1 m²

2. Stephanie; Sample answer: Estimate with 2 * 7 = 14 to get 12.87.

3. Sample answer: 3 * 3 = 9; 11.186

4. Sample answer: 70 * 5 = 350; 336.48

5. Sample answer: 7 * 60 = 420; 7 * $56.45; $395.15

6. 4 of 0.05; 0.5 * 4 is 2, and that is too much.

Home Link 3-5

1. 25 R16; 38 * 25 + 16 = 966

2. 138 R4; 43 * 138 + 4 = 5,938

3. Missing digits: 28, 12, 3, 5

4. Missing digits: 3, 204, 30, 306

5. About 32   6. About 165

Home Link 3-6

1. **a.** 56.4; 56.4 * 3.9 = 219.96
   b. 6.87; 6.87 * 0.52 = 3.5724
   c. 0.345; 0.345 * 6.8 = 2.346
   d. 0.87; 0.87 * 1.95 = 1.6965

2. 8.3; 8.3 * 0.72 = 5.976   **3.** 4.82; 4.82 * 1.6 = 7.712

4. 3 ÷ 0.4 = 7.5; 7 parfaits; 7.5 * 0.4 = 3

Home Link 3-7

1. 78 in.

2. 3 yd; Sample explanation: 78 inches is $6\frac{1}{2}$ feet. Cloth is sold in yards, so she needs 3 yards.

Home Link 3-8

1. 0.63, $\frac{63}{100}$; 0.32, $\frac{32}{100}$; 0.05, $\frac{5}{100}$

2. **a.** Sample answer: She only rolled 50 times, not 100 times.   **b.** 40%
   c. Sample answer: I doubled 50 and 20.

Home Link 3-9

1. **b.** $\frac{2}{10}$ **c.** 20% **d.** 40 **2. b.** $\frac{11}{20}$ **c.** 55% **d.** 33

3. **a.** 6   **b.** 20
   c. Sample answer: Since 25% is equivalent to $\frac{1}{4}$, I multiplied 80 by $\frac{1}{4}$ to get 20.

Home Link 3-10

1. $\frac{35}{100}$, 0.35, 35%; $\frac{52}{100}$, 0.52, 52%; $\frac{12}{30}$ (or $\frac{40}{100}$), 0.4, 40%

2. 160 pages   **3.** 120 loaves

Home Link 3-11

1. 40 **2.** 320 cm **3.** Sample answer: About 97%

4. Sample answer: About 18% **5.** 4,500 tigers

Home Link 3-12

1. Quartiles   **2.** Sample answer: How many data points there are   **3.** 4.5 cm

4. Sample answer: The rose leaves are shortest. Almost $\frac{3}{4}$ of the rose leaves are shorter than the hawthorn and juniper leaves.

5. Hawthorn

Home Link 3-13

1. Gold: 1, 6, 7, 13, 46; Silver: 1, 5, 9, 17, 29; Bronze: 4, 6, 11, 17, 32

2.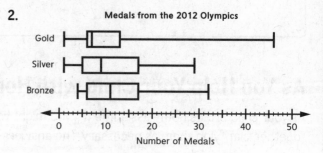

3. Gold: 7; Silver: 12; Bronze: 11

4. Sample answer: Half the countries with more than 12 won between 6 and 13 gold medals.

Home Link 3-14

1. 1   **2.** 2

4. Sample answer: There is one really tall bar on Histogram A, so the IQR (or box) would be shorter.

5. Sample answer: The data is more balanced, with about half the data below 50 and half above 50.

Reviewing Place Value with Decimals

① Record two decimals that are equivalent to each decimal below.

a. 0.2 _____

b. 0.13 _____

c. 2.145 _____

d. 7.06 _____

② Compare using <, >, or =.

a. 0.05 ____ 0.050   **b.** 0.05 ____ 0.5   **c.** 0.503 ____ 0.53

③ Explain why the zeros are necessary in 10.03 but not in 0.350.

_____

_____

_____

_____

④ Circle the numbers that are equivalent.

0.21   0.021   0.201   0.210   0.0021

⑤ Cross out the names that do not belong. Add two names to each box.

a.

0.23
$(2 * \frac{1}{10}) + (3 * \frac{1}{100})$
$(2 * \frac{1}{100}) + (3 * \frac{1}{1,000})$
$(2 * 0.01) + (3 * 0.001)$
$\frac{23}{100}$
$0.23 * 1$

b.

10.045
$(1 * 10) + (4 * 0.01) + (5 * 0.001)$
$\frac{10,045}{1,000}$
$\frac{10,045}{10,000}$
$10,045 * 0.01$
$10,045 * 1$

Practice

⑥ $6 \div \frac{1}{2} =$ _____

⑦ $2 \div \frac{1}{4} =$ _____

⑧ $5 \div \frac{1}{3} =$ _____

Decimals on the Number Line

Fido the flea is at it again. He starts at 0 and wants to go to the Flea Fair at 0.28 on the number line. Hop Set 1 takes a total of 10 hops to reach 0.28. Hop Set 2 takes a total of 28 hops to reach 0.28. Remember that the size of Fido's hops are always 1 tenth, 1 hundredth, or 1 thousandth.

SRB
112

① Show the two different hop sets on the number lines below.

Hop Set 1:

Hop Set 2:

② Write a number sentence to represent each hop set to 0.28.

Hop Set 1: _____

Hop Set 2: _____

③ **a.** Write 3.48 in expanded form as the sum of multiplication with decimals.

_____

b. Write a number between 3.48 and 3.49. _____

c. Explain how the expanded form of the number you wrote for Part b would be similar to the expanded form of 3.48 you recorded for Part a.

_____

_____

④ Circle the numbers below that are between 8.032 and 8.033.

8.035   8.03024   8.0323   8.0329   8.0335   8.032222

Practice

Insert <, >, or = to make each number sentence true.

⑤ 3.4 ____ 3.40   ⑥ 17.062 ____ 17.006   ⑦ 12.405 ____ 12.41

69

Great Accomplishments in Sports

SRB
118-128

① Geoffrey Mutai (Kenya) set the record for the New York City Marathon
in 2011. His time was 2 hours, 5.10 minutes.
In 2013, he won the marathon again with a time of 2 hours, 8.40 minutes.

How much faster was Mutai's
time in 2011 than in 2013? _____

② At the 1908 Olympics, Erik Lemming (Sweden) won the javelin throw. He threw the
javelin 54.82 meters. He won again in 1912 with a throw of 60.64 meters.

How much longer was his
1912 throw than his 1908 throw? _____

③ At the 1984 Olympics, Gregory Louganis (United States) won a gold medal in men's
springboard diving.
To calculate a diver's final score, the average scores from 11 dives are added.

Dive #1	Dive #2	Dive #3	Dive #4	Dive #5	Dive #6	Dive #7	Dive #8	Dive #9	Dive #10	Dive #11
47.52	53.01	44.16	40.32	68.88	81.00	85.56	77.40	71.1	93.06	92.40

What was Louganis's winning final score? _____

④ Driver Buddy Baker (Oldsmobile, 1980) holds the record for the fastest winning speed
in the Daytona 500. His speed was 177.602 miles per hour. Bill Elliott (Ford, 1987)
has the second-fastest winning speed. Elliott's speed was 1.339 miles per hour
slower than Baker's speed.

What was Elliott's speed? _____

Practice

⑤ $\frac{4}{5} \div \frac{1}{2} =$ _____

⑥ $\frac{3}{4} \div \frac{2}{3} =$ _____

⑦ $\frac{5}{6} \div \frac{1}{4} =$ _____

⑧ $\frac{2}{3} \div \frac{3}{4} =$ _____

Decimal-Multiplication Review

Use estimation to solve Problems 1–2.

SRB
134-138

1. Carlos is building a flower bed that is 13.2 m by 6.75 m.
   When he multiplied, Carlos got 89100.
   Show where he should place the decimal point _____

2. Stephanie says 1.95 * 6.6 = 12.87.
   Dante says the answer is 128.7. Who is right? _____
   Explain how an estimate might help you decide.

   _____

For Problems 3–5, record a number sentence to show how you estimated. Then use the U.S. traditional multiplication algorithm to solve. Use your estimate to check your work.

3. 3.4 * 3.29

   Estimate: _____

   Answer: _____

4. 70.1 * 4.8

   Estimate: _____

   Answer: _____

5. Mr. Murphy is building a fence. He bought 7 packages of wooden fencing.
   One package costs $56.45. How much do they cost all together?

   Estimate: _____

   Number model: _____   Solution: _____

Try This

6. Dr. Goode prescribes 0.2 gram of cold medicine for Donald. This medicine comes in
   tablets that are 0.05 gram or 0.5 gram. Should Donald take 4 of the 0.5 gram

   tablets or 4 of the 0.05 gram tablets? _____

   How do you know? _____

Practice

Compare with >, <, or =.

7. −7 ___ −3   8. 4 ___ −4   9. 0 ___ −3   10. −2 ___ −5

73

Long Division

Solve each problem.
Write a number sentence to show how you checked your answer.

① 38)966

② 43)5,938

Check: _____

Check: _____

Fill in the missing numbers.

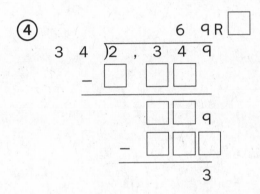

③
```
           8  3 R5
    4  1 )3 , 4  0  8
      -  3  □□
         ____
         □□    8
      -  1  2  □
         ____
            □
```

④
```
              6  9 R □
    3  4 )2 , □  3  4  9
      -  □  □□
         ____
            □□    9
         -  □□□
            ____
               3
```

⑤ There are about 1,575 beads in a large economy-size tub at the craft store.
There are 49 different colors.
If the colors are distributed equally, about how
many beads of each color are there?   _____

⑥ The book *The Phantom Tollbooth* by Norton Juster (Random House, 1961)
has 42,156 words. It is 256 pages long.
On average, how many words are on each page?   _____

Practice

⑦ ____ $= \frac{3}{4} * \frac{2}{3}$ **⑧** $\frac{4}{5} * \frac{1}{8} =$ ____   **⑨** $9\frac{2}{7} * \frac{3}{5} =$ ____   **⑩** ____ $= \frac{1}{3} * \frac{2}{9}$

75

Decimal Division

(1) Put the decimal point in the correct position in each quotient.
Use multiplication to check your answer.

SRB
151-154

a. 219.96 ÷ 3.9 = 5 6 · 4   Check: _____

b. 3.5724 ÷ 0.52 = 6 8 7   Check: _____

c. 2.346 ÷ 6.8 = 3 4 5   Check: _____

d. 1.6965 ÷ 1.95 = 0 8 7   Check: _____

Divide and check.

(2) 0.72)‾5.976‾   (3) 1.6)‾7.712‾

Check: _____   Check: _____

(4) Jaime has 3 cups of berries. Each fruit-and-yogurt parfait he makes
contains 0.4 cup of berries. How many parfaits can he make?

Number sentence: _____   Solution: _____

Check: _____

Practice

(5) GCF (10, 3) = _____   (6) GCF (12, 24) = _____

(7) GCF (100, 80) = _____   (8) GCF (18, 42) = _____

Decimal Operations

Margaret is making a pair of purple pajama pants for her daughter Marie.

To figure out how much purple fabric she needs, Margaret must do the following:

- Measure the length from Marie's waist to her ankle.

- Double this measurement.

- Add 12 inches.

(1) From waist to ankle, Marie measures 33 inches.

How many inches of the purple fabric does Margaret need? _____

(2) Cloth is sold in yards. How many yards of purple fabric will Margaret buy?
Explain why your answer makes sense.

_____

_____

_____

(3) The purple fabric costs $5.50 per yard. The tax added to Margaret's bill is $1.23.

How much does Margaret spend on the fabric? _____
Show your work.

(4) Margaret pays with a $20 bill. How much change does she receive? _____

Bring in examples of how percents are used in the world around us. You can write down, cut out, or print examples from newspapers, television, the Internet, and so on. We will collect these in a Percent Museum.

Practice

Find the LCM.

(5) LCM (8, 12) = _____

(6) LCM (4, 14) = _____

(7) LCM (10, 15) = _____

(8) LCM (9, 12) = _____

Shading Percents

(1) A recent survey investigated whether Summit Middle School students prefer to wear school uniforms. Here are the results:

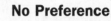

- 63 percent prefer school uniforms.
- 32 percent do not prefer school uniforms.
- 5 percent do not have a preference.

Shade each percent on the grids below. Record the decimal and fraction equivalents.

Prefer Uniforms	**Do NOT Prefer Uniforms**	**No Preference**
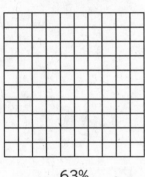	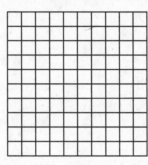	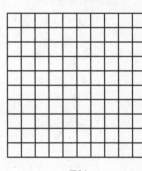
63%	32%	5%

Decimal: _____   Decimal: _____   Decimal: _____

Fraction: _____   Fraction: _____   Fraction: _____

(2) Teresa was designing a game to play at lunchtime with her friends. She wanted to know which number on a die is the luckiest. She rolled a die 50 times. The die landed showing the number five 20 times. She claimed she rolled a five 20% of the time.

a. Explain her mistake. _____

_____

b. For what percent of her 50 rolls did she roll a five? _____

c. How did you get your answer for Part b?

_____

Practice

(3) 14.7 − 13.2 = _____

(4) 4.52 − 3.5 = _____

(5) 1.2 − 0.006 = _____

(6) 3.424 − 3.006 = _____

81

Solving Percent Problems

① **a.** Shade in the grid to represent that 2 out of every 10 moviegoers buy their tickets ahead of time.

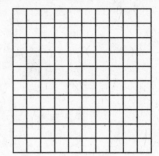

SRB
56-58

b. What fraction of moviegoers buy their tickets ahead of time? _____

c. What percent of moviegoers buy their tickets ahead of time? _____

d. If 200 people go to the movies, how many would buy their tickets ahead of time? _____

② **a.** Shade in the grid to represent that 11 out of every 20 people prefer watching movies at home instead of watching them at the theater.

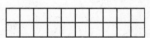

b. What fraction of people prefer to watch movies at home? _____

c. What percent of people prefer to watch movies at home? _____

d. If 60 people are asked, how many prefer to watch movies at home? _____

③ **a.** 10% of 60: _____   **b.** 25% of 80: _____

c. Explain how you found the answer to Part b.

_____

_____

④ **a.** Write $\frac{9}{10}$ as a percent. _____   **b.** Write $\frac{2}{5}$ as a percent. _____

Practice

Find the median.

⑤ 109, 121, 134, 115, 146 _____   ⑥ 11, 17, 22, 13, 35, 27 _____

Percents as Ratios

(1) Fill in the missing numbers and shade the grid.

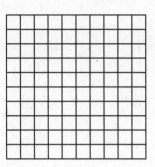

Fraction: _____

Decimal: _____

Percent: _____%

Ratio: 35 : 100

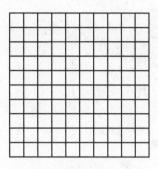

Fraction: _____

Decimal: _____

Percent: _____%

Ratio: 52 : 100

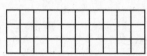

Fraction: _____

Decimal: _____

Percent: _____%

Ratio: 40 : 100

Use ratio/rate tables to solve each problem.

(2) Kiese has read 80% of his library book.
The book has 200 pages.
How many pages has he read?

_____

(3) A bakery donated 30 loaves of
bread to a homeless shelter.
That was 25% of the loaves
they made that morning.
How many loaves did they
make that morning?

_____

Practice

Write an equivalent ratio.

(4) 2 : 3 _____

(5) 5 : 6 _____

(6) 3 : 9 _____

(7) 14 : 20 _____

85

Tiger Facts

Solve.

(1) Tigers have a hunting success rate of about 10%.
A tiger successfully hunts 4 times in one week.
How many attempts did the tiger make?   _____

(2) A Bengal tiger's tail is around 30% of its total length.
The total length of one Bengal tiger's tail is 96 cm.
Around how long is the tiger?   _____

(3) At the start of the 20th century, there were about 100,000 tigers in the wild.
In 2014, there were about 3,200.
By about what percent did the tiger
population decrease?   _____

(4) Tiger cubs are around 2 years old when they leave their mothers.
In the wild, tigers live about 11 years.
About what percent of their lives
do tigers spend with their mothers?   _____

Try This

(5) About 5,000 tigers live in captivity in the United States.
About 10% of these tigers live in reputable zoos.
Around how many of these tigers DO NOT
live in reputable zoos?   _____

Practice

Compare using >, <, or =.

(6) 2.58 _____ 2.576

(7) $\frac{5}{6}$ _____ $\frac{8}{9}$

(8) $\frac{7}{8}$ _____ 0.875

(9) $\frac{4}{7}$ _____ 0.59

87

Box Plots

SRB
301-302

Fill in the blanks about a five-number summary you could use to make a box plot.

①   These five numbers divide the data into four _____.

②   What can you NOT tell from a box plot? _____

_____

Use the box plot to answer the questions in Problems 3–5.

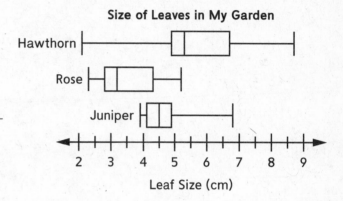

Size of Leaves in My Garden

Leaf Size (cm)

③   Half of the juniper leaves are longer than what measurement? _____

④   Which plant has the shortest leaves?

How do you know?

_____

_____

⑤   Which type of leaf varies the most in length? _____

Use the box plot to answer the questions in Problems 6–7.

2009 Attendance at MLB Stadiums

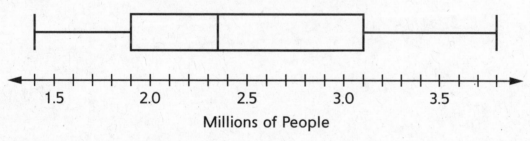

Millions of People

⑥   The middle 50% of attendance at MLB stadiums is between _____ and _____ million people.

⑦   Which quarter of the data has the greatest range? _____

Practice

⑧   If 50% of a number is 14, then 100% of the number is _____.

⑨   If 10% of a number is 6, then 100% of the number is _____.

89

Box Plots for Olympic Medals

SRB
301-302

Countries often win more than one medal at the Olympic games.
Nineteen countries won more than 12 medals each at the London Olympic games
in 2012. Listed below are the numbers of medals won by each of those countries.

- Gold: 1, 3, 3, 5, 6, 6, 6, 7, 7, 7, 8, 8, 11, 11, 13, 24, 29, 38, 46

- Silver: 1, 2, 3, 4, 5, 5, 5, 6, 8, 9, 10, 11, 14, 16, 17, 19, 26, 27, 29

- Bronze: 4, 5, 5, 5, 6, 7, 8, 9, 9, 11, 12, 12, 12, 14, 17, 19, 23, 29, 32

(1) List the five-number summary for each type of medal.

Gold: _____   Silver: _____

Bronze: _____

(2) Make a box plot for
each type of medal:
gold, silver, and bronze.
Make all three box plots,
one above the other, on
the number line at right.

Medals from the 2012 Olympics

Gold

Silver

Bronze

0   10   20   30   40   50

Number of Medals

(3) List the IQR for each type of medal.

Gold: _____   Silver: _____   Bronze: _____

(4) What does the IQR tell you about the number of gold medals that were won?

_____

_____

Practice

Find the equivalent unit ratio.

(5) 4 : 8 _____

(6) 5 : 15 _____

(7) 66 : 33 _____

(8) 56 : 14 _____

Matching Histograms and Box Plots

Below are two histograms and two box plots.

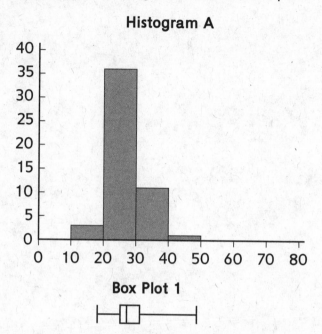

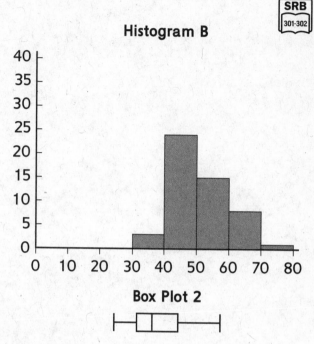

(1) Box Plot _____ matches Histogram A.   (2) Box Plot _____ matches Histogram B.

(3) Sketch each box plot above its corresponding histogram.

(4) Explain how you know which box plot matches the data shown in Histogram A.

_____

_____

(5) Explain how you know which box plot matches the data shown in Histogram B.

_____

_____

Try This

(6) The title Median Family Income by State (in thousands) matches Histogram _____.

(7) The title Percent of Adults with College Degrees by State matches Histogram _____.

Practice

Divide.

(8) $4.2 \div 2.1 =$ _____   (9) $36 \div 0.6 =$ _____   (10) $0.15 \div 0.05 =$ _____

93

Algebraic Expressions and Equations

Unit 4 begins with a review of exponential notation and order of operations. These topics are integral to algebra, which is introduced in this unit.

To develop algebraic thinking, students work with numerical expressions and number sentences. They use expressions with variables to write simple equations and to generalize patterns. Students apply properties of numbers and operations and the order of operations to identify and generate equivalent expressions.

In several lessons in this unit, students extend their work with number sentences to inequalities, comparing two quantities that are not equal. They learn how to write and graph inequalities.

Students apply these algebraic concepts to real-world situations. For example, amusement park rides sometimes have a minimum height requirement of 4 feet. In real life, an individual can be only so tall, so there is also a probable maximum height. In this example, a rider must be 4 feet tall, and we could consider the maximum height for a person to be 9 feet. Students first define a variable to work with: let h represent the height of a rider in feet. Then they write and graph inequalities that model this situation: $4 \leq h$ and $h \leq 9$.

The unit concludes with lessons on absolute value and the uses of absolute value. Students explore the absolute value of a number as its distance from 0 on a number line. Absolute value emphasizes magnitude (distance) rather than whether the number is negative or positive. This emphasis is important in real-life contexts. For example, students want to compare estimates for the number of jelly beans in a jar. Some estimates will be too high and some too low. To evaluate whose estimate is closest, students compare the distances the estimates are from the actual total, disregarding whether a guess is above or below the actual total. This situation could be represented using absolute value.

One focus of the absolute-value lessons is to help students make distinctions between using integers to represent situations and using absolute value. For example, a balance of −$50 in a bank account indicates that the owner of the bank account owes the bank $50. Students use an integer to represent the bank balance. If they want to answer the question of how much money is owed, one approach is to consider the distance the balance is from 0. They can use $|-50|$, or the absolute value of −50, to represent the amount of money that is owed, or the total amount of the debt.

Please keep this Family Letter for reference as your child works through Unit 4.

Vocabulary

Important terms in Unit 4:

algebraic expression An expression that contains one or more unknowns or variables. For example, if Maria is 2 inches taller than Oliver, and if the variable a represents Maria's unknown height, then the algebraic expression $a - 2$ represents Oliver's height.

Distributive Property of Multiplication over Addition A property that relates multiplication of a sum of numbers by distributing the factor over the terms in the sum.

Example: $2 * (5 + 3) = (2 * 5) + (2 * 3) = 10 + 6 = 16$

In symbols: For any numbers a, b, and c, $a * (b + c) = (a * b) + (a * c)$.

Distributive Property of Multiplication over Subtraction A property that relates multiplication of the difference of numbers by distributing the factor over the terms in the difference.

Example: $2 * (5 - 3) = (2 * 5) - (2 * 3) = 10 - 6 = 4$

In symbols: For any numbers a, b, and c, $a * (b - c) = (a * b) - (a * c)$.

equation A number sentence with an $=$. For example, $15 = 10 + 5$ is an equation.

exponential notation A way to show repeated multiplication by the same factor. For example, 2^3 is exponential notation for $2 * 2 * 2$. The 3 is the *exponent*. It tells how many times the number 2, called the *base*, is used as a factor. You read this number as *two to the third power*.

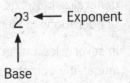

inequality A number sentence with a relation symbol other than $=$, such as $>$, $<$, $\geq$, $\leq$, $\neq$, or $\approx$.

order of operations A set of rules that tells the order in which operations in an expression are performed. The accepted order is:

1. Follow Steps 2–4 for expressions inside parentheses or grouping symbols. Work from the innermost to the outermost grouping symbols.

2. Calculate all expressions with exponents.

3. Multiply and divide in order from left to right.

4. Add and subtract in order from left to right.

Example 1:	Example 2:
$10 - 4 + 3 = 6 + 3$ Subtract (left to right).	$5^2 + (3 * 4 - 2) / 5$ Calculate inside the parentheses.
$\quad\quad\quad = 9$ Add.	$= 5^2 + (12 - 2) / 5$
	$= 5^2 + 10 / 5$ Calculate inside the parentheses.
	$= 25 + 10 / 5$ Simplify the exponential expression.
	$= 25 + 2$ Divide.
	$= 27$ Add.

variable A letter or symbol that can be replaced by a value. A variable can represent one particular number or many different numbers.

Do-Anytime Activities

Work with your child on some of the concepts taught in Unit 4:

1. Ask your child how to play the game *Name That Number*. Play a few rounds of the game several times during this unit. You can play the game online if your child's class has access to the *Everyday Math* online games. If not, you can use a regular deck of 54 playing cards as follows:

   • Let all number cards represent their face values.

   • Let the four aces be 1s.

   • Change the four queens to 0s.

   • Remove the four jacks, four kings, and two jokers. Label each of these ten cards with one of the numbers from 11 through 20.

2. Think of a number between 25 and 100 and ask your child to guess it by asking questions that involve inequalities: less than, greater than, less than or equal to, or greater than or equal to. Encourage them to use operations when guessing. For example, your child could ask questions such as "Is 10 times your number greater than 500? Is 25 less than your number greater than 10?" Your child should keep asking questions until he or she figures out the number. The goal is to find the number using as few questions as possible. Keep track of the number of questions it takes to figure out a number, and then switch roles with your child.

3. Ask your child to give the magnitude and direction for situations that have numbers below 0. For example, a temperature reading of −10°F means that the temperature is below 0 and its magnitude is 10.

Building Skills through Games

In Unit 4, your child will work on a variety of skills by playing the following games:

Ratio Memory Match
Students match ratio picture cards to ratio notation cards.

Name That Number
Students use order of operations in number sentences to make a target number.

Fraction Top-It (Division)
Students divide whole numbers and fractions by fractions. Then they compare fractions.

Doggone Decimal
Students estimate products of whole numbers and decimals.

Solution Search
Students find solutions to inequalities.

Absolute Value Sprint
Students explore absolute value of numbers on a number line.

Percent Spin
Students calculate percents.

As You Help Your Child with Homework

As your child brings assignments home, you might want to go over the instructions together, clarifying as necessary. The answers listed below will guide you through the unit's Home Links.

Home Link 4-1
2. **a.** 144   **b.** 64   **c.** 18   **d.** 12

3. $3 * 3 * 3$, 27; 6^2, 36; 7^4, 2,401

Home Link 4-2
4. **a.** 23   **b.** 2,916

5. Sample answers: $1 = 7 \div 7 * \frac{7}{7} * \frac{7}{7}$; $6 = 7 - \left[\frac{(7 + 7)}{7} - \frac{7}{7}\right]$

Home Link 4-3
3. Sample answer: $2 * (13 + 1)$; 28

4. Expressions equivalent to $2 * n + 2$; 252 tiles

Home Link 4-4

2. **a.** h; the height of the rectangle

   b. Sample answers: $h \div 2$, $\frac{h}{2}$, or $\frac{1}{2}h$

   c. Sample answers: $h + h + (h \div 2) + (h \div 2)$, or $3h$

4. **a.** $4n$ **b.** $n \div 4$ **c.** $2n + 4$

Home Link 4-5

2. **a.** Many   **b.** One   **c.** None

3. **a.** Always true   **b.** Cannot tell

4. Sample answer: It is sometimes true, depending on what you substitute for t.

Home Link 4-6

1. **a.** K **b.** L **c.** J **d.** J **e.** K **f.** L **g.** L **h.** J

2. a. $(80 * 5) + (120 * 5) = (80 + 120) * 5$ and
   c. $\left(9 * \frac{3}{8}\right) - \left(\frac{2}{3} * \frac{3}{8}\right) = \left(9 - \frac{2}{3}\right) * \frac{3}{8}$

Home Link 4-7

2. **a.** False; None   **b.** False; None

   c. True; Commutative Property of Multiplication

3. **a.** $85 * (100 + 1) = 85 * 100 + 85 * 1 = 8{,}500 + 85 = 8{,}585$

4. **a.** $12 * (4 + 2) = 12 * 6$

Home Link 4-8

1. 22 toothpicks   2. 42 toothpicks

3. Sample answer: The first triangle is always made of 6 toothpicks. You add on 4 toothpicks each time.

5. 20

Home Link 4-9

1. Sample answer: Let z be the number of text messages; $z \leq 500$

4. **a.** $x < 42$ **b.** $x > 42$ **c.** $x \geq 42$ **d.** $x \leq 42$

Home Link 4-10

1. **a.**

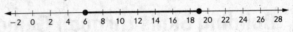

   b.

2. **a.** $-1 > x$ (or $x < -1$)   **b.** $x \leq 4$ (or $4 \geq x$)

   c. Sample answer: -10, -2.5, $-1\frac{1}{4}$

3. **a.** $y < 0$ **b.** Sample answer: 1, 5, 8

   c. Sample answer: $y > 8$

Home Link 4-11

3. Let n be the number of feet; $n \leq 19$ and $n \geq 6$.

4. Let g be the number of eggs; $g \leq 50$ and $g \geq 20$.

Home Link 4-12

2. **a.** 20   **b.** 8.25   **c.** 79

   d. 0.004   **e.** $10\frac{1}{2}$ **f.** 0

3. **a.** Any positive number or 0

   b. Any negative number or 0

Home Link 4-13

2.

Jamal; $|-18| > |-14|$

4. **a.** 4   **b.** 5   **c.** 3.5   **d.** 41

Home Link 4-14

1. Sample answer: It will have a small mean absolute deviation because it is close to the ocean, where temperatures are more consistent.

2. **a.** 41 **b.** 66 **c.** 51.5 **d.** 52.3 **e.** 25 **f.** 8

4. 3   5. 2.5   6. 50   7. 3.25

Using Order
of Operations

(1) Insert parentheses to make the expression equivalent to the target number.

SRB
98

Numerical Expression	Target Number
8 − 2 + 5	1
15 − 3 * 4 + 2	50
3 * 5 + 4 * 6	162

(2) Simplify each expression.

a. $(3 + 9)^2$ _____

b. $2^4 * 2^2$ _____

c. $20 − (6 − 4)$ _____

d. $\left(\frac{1}{2} \div \frac{1}{4}\right) * 6$ _____

(3) Complete the table.

Exponential Notation	Multiplication Expression	Standard Notation
4^2	4 * 4	16
3^3		
	6 * 6	
	7 * 7 * 7 * 7	

(4) Use the given calculator keys to find an expression equivalent to the target number. You may use the keys more than once or not at all.

Keys	Target	Expression
③ ② (∧) (×) (+) (Enter ⚌)	29	
⑦ ③ (∧) (×) (Enter ⚌)	343	
Try This (−) ② (↓) ⑨ (Enter ⚌)	0.2222 . . .	

Practice

Write the opposite of each number.

(5) 12 _____ (6) −2 _____ (7) −3.5 _____ (8) $\frac{3}{5}$ _____

99

Practicing Order of Operations

In Problems 1–3, tell whether the number sentence is true or false. If it is false, rewrite it with parentheses to make it true.

SRB
203

Number Sentence	True or False	Correction (If Needed)
① $4 + 8 \div 4 + 4 = 5$	_____	_____
② $46 = 3 * 6 + 7 * 4$	_____	_____
③ $15 - 12 \div 3 + 6 \div 2 = 8$	_____	_____

④ Evaluate.

a. $45 - (1 + 4)^2 + 3$ **b.** $(2 + 4)^2 * (1 + 2)^4$

⑤ Write an expression for AT LEAST three of the following numbers using six 7s. All values can be found using only addition, subtraction, multiplication, and division.

1 = _____

2 = _____

3 = _____

4 = _____

5 = _____

6 = _____

Practice

Find the greatest common factor.

⑥ GCF (10, 50) = ____   **⑦** GCF (80, 24) = ____   **⑧** GCF (90, 54) = ____

Using Expressions

(1) **a.** Write a numerical expression for calculating the number of shaded border tiles for the pictured 12-by-12 tiled floor.

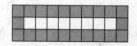

SRB
225

_____

Number of shaded tiles: _____

b. Circle the expressions below that also represent the number of shaded tiles in the 12-by-12 tiled floor.

11 + 11 + 11 + 11   4 * 12 + 4   (12 − 2) + (12 − 2) + 12 + 12   4 * 12 − 2

c. Choose one of the expressions you circled in Part b and explain how it represents the number of shaded tiles.

_____

_____

_____

(2) A rectangular tiled floor is shown at the right. Write an expression that models how you can find the number of shaded tiles in the 3-by-10 rectangular floor.

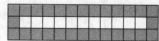

_____   Number of shaded tiles: _____

(3) Write an expression that models how you can find the number of shaded tiles in the 3-by-13 rectangular floor shown at the right.

_____   Number of shaded tiles: _____

Try This

(4) Write an algebraic expression for the number of shaded tiles in a 3-by-*n* rectangular floor. Use your expression to find the number of shaded tiles in a 3-by-125 tiled floor.

_____

Practice   Find the least common multiple.

(5) LCM (3, 5) = ____   **(6)** LCM (10, 12) = ____   **(7)** LCM (6, 12) = ____

Algebraic Expressions

Write an algebraic expression. Use your expression to solve the problem.

SRB
212-214

(1) Kayla has *x* hats. Miriam has 6 fewer hats than Kayla. _____

If Kayla has 22 hats, how many hats does Miriam have? _____

(2) The width of Rectangle A is half of its height.
Write an algebraic expression for the width of Rectangle A.

a. Define your variable. Let ____ represent _____.

b. Algebraic expression: _____

c. Using the variable you defined in Part a, write an algebraic expression for the perimeter of Rectangle A. _____

(3) Larry ran 2.5 miles more than Jusef.
Write an algebraic expression for how far Larry ran.

a. Define your variable. Let ____ represent _____.

b. Algebraic expression: _____

c. If Jusef ran 5 miles, how many miles did Larry run? _____

(4) For each situation, choose an expression from the box that matches the situation, and write it in the matching blank. You may use an expression more than once.

$n \div 25$	$2n + 4$	$n \div 4$	$4n + 2$
$4n$	$n - 4$	$n + 4$	$25 \div n$

a. With 4 bags of *n* potatoes, the total number of potatoes is _____.

b. If you exchange *n* quarters for dollars, you get _____ dollars.

c. There are *n* pens in a box. Denise has 4 pens more than 2 boxes of pens.

The total number of pens Denise has is _____.

Practice

Use <, >, or = to make the number sentence true.

(5) $\frac{3}{4}$ ____ $\frac{3}{7}$ (6) 0.4 ____ 0.400   (7) 0.8 ____ 0.67

105

Equations

SRB
226-227

① Look for a pattern in the set of numerical equations. Describe the pattern in words. Use a variable and write an equation that represents the pattern.

$$3^6 = 3^2 * 3^4 \qquad 58^6 = 58^2 * 58^4 \qquad (0.25)^6 = (0.25)^2 * (0.25)^4$$

a. Description: _____

_____

b. Equation that generalizes the pattern: _____

c. Write two more examples of the pattern: _____

_____

② For each equation, circle the number of solutions you could find.

    a. $c + c = 2 * c$ Many   None   One

    b. $28 = t - 2$ Many   None   One

    c. $m - 1 = m - 2$ Many   None   One

③ Circle the answer that best describes each equation.

    a. $c + c = 2 * c$ Always true   Never true   Cannot tell

    b. $28 = t - 2$ Always true   Never true   Cannot tell

④ Explain your answer to Problem 3b. _____

_____

Try This

⑤ The numbers 4, 5, and 6 are called *consecutive numbers* because they follow each other in order. The sum of 4, 5, and 6 is 15—that is, $4 + 5 + 6 = 15$. Circle all equations that generalize finding a sum of 170 for three consecutive numbers.

    a. $x + 2x + 3x = 170$ **b.** $170 = x + (x + 1) + (x + 2)$ **c.** $3x + 3 = 170$

Practice   Estimate whether each sum is closest to 0, $\frac{1}{2}$, 1, or $1\frac{1}{2}$.

⑥ $\frac{8}{9} + \frac{5}{8}$ _____

⑦ $\frac{1}{10} + \frac{1}{11}$ _____

⑧ $\frac{5}{6} + \frac{2}{16}$ _____

The Distributive Property

(1) Each of the expressions describes the area of the shaded part of one of the rectangles. Write the letter of the correct rectangle next to each expression.

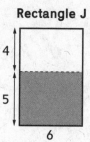
Rectangle J

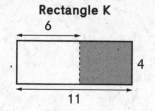
Rectangle K

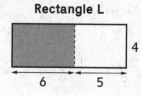
Rectangle L

a. $4 * (11 - 6)$ _____

b. $44 - 20$ _____

c. 30 _____

d. $(6 * 9) - (6 * 4)$ _____

e. $(4 * 11) - (4 * 6)$ _____

f. $(11 - 5) * 4$ _____

g. $(11 * 4) - (5 * 4)$ _____

h. $6 * (9 - 4)$ _____

(2) Circle the equations that are examples of the Distributive Property.

a. $(80 * 5) + (120 * 5) = (80 + 120) * 5$

b. $6 * (3 - 0.5) = (6 * 3) - 0.5$

c. $\left(9 * \frac{3}{8}\right) - \left(\frac{2}{3} * \frac{3}{8}\right) = \left(9 - \frac{2}{3}\right) * \frac{3}{8}$

d. $(16 * 4) + 12 = (16 + 12) * (4 + 12)$

Write an equation to show how the Distributive Property can help you solve each problem.

(3) Kelly signed copies of her new book at a local bookstore.
In the morning she signed 36 books, and in the afternoon she signed 51 books.
It took her 5 minutes to sign a book. How much time did she spend signing books?

Equation: _____ Solution: _____

(4) Mr. Katz gave a party because all the students scored 100% on their math tests.
He had budgeted $1.15 per student. It turned out that he spent $0.25 less per student. How much money did he spend for 30 students?

Equation: _____ Solution: _____

Practice Write the reciprocal.

(5) 5 _____

(6) $\frac{2}{9}$ _____

(7) $3\frac{1}{3}$ _____

Applying the Distributive Property

SRB
204-205,
231-232

(1) Match each property with a generalized form of the property.

Commutative Property of Addition $a * (b * c) = (a * b) * c$

Commutative Property of Multiplication $a * b = b * a$

Associative Property of Addition $a * (b + c) = ab + ac$

Associative Property of Multiplication $(a * b) - (a * c) = a * (b - c)$

Distributive Property of Multiplication $a + b = b + a$
    over Addition $(a + b) + c = a + (b + c)$

Distributive Property of Multiplication
    over Subtraction

(2) For each equation below, use general equations for properties to determine whether it is true or false. For each true number sentence, list the property or properties that apply. For false number sentences, write "None."

a. $(9 - 4) * 3 = (9 - 3) * (4 - 3)$ _____ Property: _____

b. $(8 + 5) * 2 = (8 + 2) * (5 + 2)$ _____ Property: _____

c. $(8 + 5) * 2 = 2 * (8 + 5)$ _____ Property: _____

Use the Distributive Property to solve Problems 3–4.

(3) Show how to solve the problems mentally.

a. $85 * 101 = $ _____

b. $156 * 9 = $ _____

c. $48 * 24 = $ _____

(4) Rewrite each expression as a product by taking out a common factor.

a. $48 + 24 = $ _____ * (_____ + _____) = _____ * _____

b. $72 - 56 = $ _____ * (_____ - _____) = _____ * _____

c. $(2y) + (3 * y) = $ (_____ + _____) * _____ = _____ * _____

Practice

Use <, >, or = to make the sentence true.

(5) $\frac{2}{3}$ ___ $\frac{2}{5}$ **(6)** 0.7 ___ $\frac{4}{5}$ **(7)** 0.3 ___ 0.23 **(8)** $1\frac{1}{4}$ ___ 1.25

Building with Toothpicks

Yaneli is building a pattern with toothpicks. The pattern grows in the following way:

Design 1

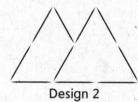
Design 2

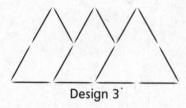
Design 3

SRB
225

① How many toothpicks are needed for Design 5? _____

② How many toothpicks are needed for Design 10? _____

③ Describe in words how you see the toothpick design growing. What stays the same from one figure to the next? What changes?

_____

_____

④ Write an expression to represent how many toothpicks are needed for Design *n*?

_____

⑤ What toothpick design number could you build with exactly 82 toothpicks? _____

⑥ Describe how you can figure out the number of toothpicks you need for any design number.

_____

_____

_____

Practice

Evaluate each expression.

⑦ $7^2 =$ _____

⑧ _____ $= 2^4$

⑨ $1^5 =$ _____

⑩ $4^3 =$ _____

Inequalities

(1) Amelia's cell phone plan lets her send a maximum of 500 text messages per month.

Define a variable. _____

Write an inequality to represent Amelia's situation. _____

(2) The temperature in the freezer should be no higher than −18°C.

Define a variable. _____

Write an inequality to represent the situation. _____

(3) Sam scored 68 in miniature golf. What score would beat Sam's score?

Define a variable: _____

Write an inequality to represent the situation. _____

(4) Choose the number sentence that represents each statement.

$x \geq 42$	$x > 42$
$x \leq 42$	$x < 42$

    a. A number is less than 42. _____

    b. A number is greater than 42. _____

    c. A number is at least 42. _____

    d. A number is no greater than 42. _____

Practice

(5) _____ $= 5.6 + 11.7$ **(6)** $9.2 +$ _____ $= 12.1$

(7) $19.37 − 9.29 =$ _____   **(8)** _____ $= 0.834 − 0.75$

Solving and Graphing Inequalities

Describe the solution set for each inequality.
Graph the solutions for each inequality.

SRB
210-211

(1) **a.** $5 < n$ _____

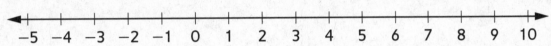

b. $q < 5$ _____

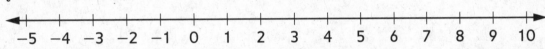

c. $w > -3$ _____

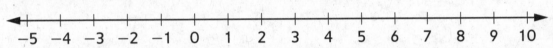

(2) Write the inequality represented by each graph below.

a. 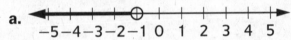   **b.**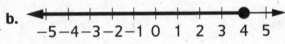

_____   _____

c. List three numbers that are part of the solution set for Part a.

_____

(3) **a.** Write an inequality with a solution set that is all numbers less than 0. _____

b. Find three numbers that are not in the solution set for Part a.

_____

c. Write an inequality with a solution set that does not have any numbers in common with the solution set in Part a or the numbers you wrote in Part b.

_____

Practice

Solve.

(4) $3.45 * 2 =$ _____   **(5)** $3.2 * 4.5 =$ _____   **(6)** _____ $= 1.53 * 3.3$

117

Graphing Alligator Facts

(1) If the temperature of an alligator nest is below 86°F, the female alligators hatch.

Define a variable: _____

Represent the statement with inequalities: _____
Graph the solution set that makes both inequalities true.

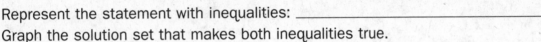

Describe how your graph represents the situation.

_____

(2) If the temperature of an alligator nest is above 93°F, the male alligators hatch.
Use the same variable you used in Problem 1.

Represent the statement with inequalities: _____
Graph the solution set that makes both inequalities true.

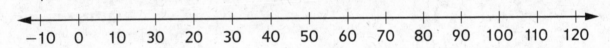

(3) Adult alligators are at least 6 feet long. The longest one on record was 19 feet.

Define a variable: _____

Represent the statement with inequalities: _____
Graph the solution set that makes both inequalities true.

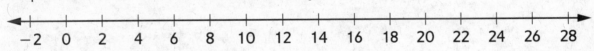

(4) Alligators lay 20–50 eggs in a clutch. Variable: _____

Represent the statement with inequalities: _____
Graph the solution set that makes both inequalities true.

Describe how your graph represents the situation.

_____

_____

Practice Evaluate.

(5) 15% of 60 _____   **(6)** 25% of 300 _____   **(7)** 250% of 18 _____

Absolute Value

(1) **a.** On the number line, plot points at two numbers whose absolute values are 8.

```
◄──┼─┼─┼─┼─┼─┼─┼─┼─┼─┼─┼─┼─┼─┼─┼─┼─┼─┼─┼─┼─►
   −10        −5         0          5         10
```

b. Explain why you get a positive number when you take the absolute value of a negative number.

_____

_____

(2) Complete.

a. $|20| =$ _____   **b.** $|8.25| =$ _____   **c.** $|-79| =$ _____

d. $|-0.004| =$ _____   **e.** $|-10\frac{1}{2}| =$ _____   **f.** $|0| =$ _____

(3) Find at least three numbers that answer each riddle.

a. A number with an absolute
value that is equal to itself _____

b. A number with an absolute
value that is its opposite _____

(4) Make up your own absolute value riddle.

_____

_____

Try This

(5) Find at least three numbers that make each statement true.

a. $|x| = -x$ _____

b. $|x| > -x$ _____

Practice   Divide. Express your remainder as a fraction.

(6) $8\overline{)3,254}$ **(7)** $52\overline{)7,859}$

121

Using Absolute Value

For Problems 1–2, do the following:

- Plot the numbers on the number line.

- Answer the question.

- Circle the number model that supports your answer.

(1) The freezing point of water is 0°C. In Chicago, it is −7°C. In Montreal, it is −9°C.

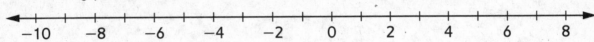

Which city's temperature is farther from 0? _____

$-7 > -9$ or $|-9| > |-7|$

(2) Rita has a debt of $14, and Jamal has a debt of $18.

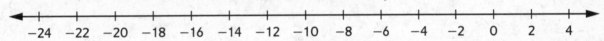

Whose balance is farther from 0? _____

$|-18| > |-14|$ or $-18 < -14$

(3) Explain how you know whether you need to use absolute value to answer the question. What do you have to consider?

_____

(4) Find the distance between the ordered pairs.

    a. $(-2, -1)$ and $(-2, 3)$ Distance: _____

    b. $(-2, 3)$ and $(3, 3)$ Distance: _____

    c. $(3, -1)$ and $(3, -4.5)$ Distance: _____

    d. $(-11, 9)$ and $(-11, -32)$ Distance: _____

Practice   Solve.

(5) $2\frac{1}{2} \div \frac{3}{4} =$ _____

(6) $1\frac{2}{3} \div \frac{1}{3} =$ _____

(7) $3\frac{3}{4} \div \frac{1}{3} =$ _____

Temperatures in Seattle

The city of Seattle is located in the state of Washington.
It is located 113 miles south of the U.S.–Canadian border at a latitude of 47°37' N.
The city is located at sea level on Puget Sound, near the Pacific Ocean.

SRB
291-294

1. Use the information above to predict whether Seattle's monthly average temperature data will have a large or small mean absolute deviation. Explain your answer.

_____

_____

2. The average monthly temperatures for Seattle are given below. Find the listed data landmarks and measures of spread. Round your answers to the nearest tenth.

Average Monthly Temperatures (°F)											
Jan	Feb	Mar	Apr	May	Jun	Jul	Aug	Sep	Oct	Nov	Dec
41	43	46	50	56	61	65	66	61	53	45	41

a. Minimum: _____   b. Maximum: _____   c. Median: _____

d. Mean: _____   e. Range: _____   f. Mean absolute deviation: _____

3. Use the data landmarks and measures of spread you found in Problem 2 to draw some conclusions about Seattle's average monthly temperatures.

_____

_____

_____

Bring in one 3-dimensional shape with faces made up of polygons. It will go in the class Shapes Museum. Find a shape that has at least one face that is not a rectangle. See pages 246–248 in your Student Reference Book for examples of the kinds of shapes to bring.

Practice   Solve.

4. _____ = 0.09 ÷ 0.03

5. 0.75 ÷ 0.3 = _____

6. 24 ÷ 0.48 = _____

7. _____ = 5.2 ÷ 1.6

Area and Volume Explorations

Unit 5 extends your child's understanding of area and volume. The unit begins with a brief review of polygons, during which students plot polygons on a coordinate grid. Then, in order to determine the lengths of the sides of these polygons, your child will apply his or her prior experience using absolute value to find distances on a coordinate grid.

Throughout the unit, your child will explore area in real-world and mathematical contexts. The lessons require your child to apply his or her work with finding the area of rectangles in earlier grades to deriving area formulas for parallelograms and triangles. Your child will also investigate how to find the areas of more complex figures by decomposing them into familiar polygons for which the area formulas are known.

Area of a Triangle
Area = $\frac{1}{2}$ of (base $*$ height)
$A = \frac{1}{2} * b * h$

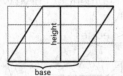

Area of a Parallelogram
Area = base $*$ height
$A = b * h$

The geometry explorations continue with 3-dimensional shapes. Your child will build geometric solids from nets and create nets for rectangular and triangular prisms.

Later lessons in the unit engage your child in extending earlier work with area and 3-dimensional shapes to finding the surface areas and volumes of solids.

Your child will then have opportunities to use the tools developed in this unit to solve fun and challenging real-world problems, such as finding the volume of a person and designing ideal shipping containers.

This unit aims to increase your child's awareness of the relevance of 2- and 3-dimensional shapes in his or her surroundings and to provide your child with situations that help him or her develop spatial reasoning.

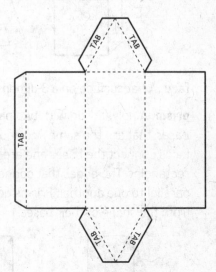

Net for a triangular prism

Please keep this Family Letter for reference as your child works through Unit 5.

Vocabulary

Important terms in Unit 5:

area The amount of surface inside a closed boundary. Area is measured in square units, such as square inches or square centimeters.

cube A polyhedron with 6 square surfaces. A cube has 8 vertices and 12 edges.

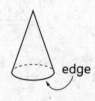

edge A line segment or curve where two surfaces meet.

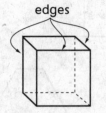

edges

edge

face A flat surface on a 3-dimensional shape.

prism A polyhedron with two parallel faces, called *bases*, that are the same size and shape. All other faces connect the bases and are shaped like rectangles. The edges that connect the bases are parallel to one another. Prisms get their names from the shape of their bases.

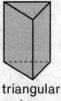

triangular
prism

rectangular
prism

surface area The total area of all the surfaces of a 3-dimensional object.

vertex (plural: vertices) The point where the sides of an angle, the sides of a polygon, or the edges of a polyhedron meet.

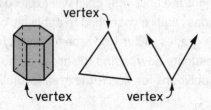

vertex

vertex

vertex

volume A measure of how much a particular container will hold. Volume is measured in cubic units, such as cubic centimeters (cm^3) or cubic inches ($in.^3$). Same as capacity, which is measured in units such as gallons or liters.

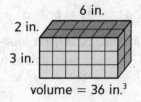

6 in.

2 in.

3 in.

volume = 36 in.^{3}

1 cubic centimeter
(actual size)

If the cubic centimeter were hollow, it would hold exactly 1 milliliter.
1 milliliter (mL) = 1 cm^3.

Do-Anytime Activities

1. Have your child compile a portfolio of 2- and 3-dimensional shapes or create a collage of labeled shapes. Collect images from newspapers, magazines, photographs, and so on.

2. Ask your child to show you how to calculate the surface areas of solids, such as tissue and cereal boxes. He or she can also help determine about how much wrapping paper you would need for different gift boxes by calculating their surface areas.

3. To explore kitchen measures, work with your child to make a favorite recipe. (Doubling the recipe can be good practice in computing with fractions.) Ask your child to use measuring cups and spoons to find the capacity of various containers. The data can be organized in a table like this one:

Container	Capacity
Coffee mug	$1\frac{1}{4}$ cups
Egg cup	3 tablespoons

Building Skills through Games

In Unit 5, your child will practice a variety of skills by playing the following games:

Polygon Capture
Students review attributes of 2-dimensional shapes as they match polygons to attributes.

Ratio Dominoes
Students identify equivalent ratios.

Multiplication Wrestling
Students multiply mixed numbers.

Name That Number
Students apply the order of operations in writing expressions equivalent to a target number.

Getting to One
Students compare decimals and use proportional reasoning to solve problems.

Ratio Comparison
Students use ratio/rate tables to match ratios to given situations.

As your child brings assignments home, you might want to go over the instructions together, clarifying them as necessary. The answers listed below will guide you through some of the Home Links in this unit.

Home Link 5-1

1. **a.** 5; (−1, −4); (1, −4)   **b.** 5; 2; (−3, 1)

2.

Horizontal Sides	Segment Endpoints	Model for Length	Length
$\overline{CD}$	(−1, −4) and (1, −4)	I−1I + I1I	2
$\overline{XY}$	(−5, 1) and (−3, 1)	I−5I − I−3I	2

Vertical Sides	Segment Endpoints	Model for Length	Length
$\overline{BC}$	(−1, 1) and (−1, −4)	I1I + I−4I	5
$\overline{AD}$	(1, 1) and (1, −4)	I1I + I−4I	5
$\overline{YZ}$	(−3, 1) and (−3, 6)	I6I − I1I	5

3. 1,221 R3   4. 45   5. 221   6. 18 R6

Home Link 5-2

1. 36 ft²   2. 24 ft²   3. 4,680 cm²   4. 28 cm²

5. 13 in.   6. 85 m   7. 9   8. 27

9. 91.3

Home Link 5-3

1. 16 ft²   2. 30 cm²   3. 6 mm²

4. 12.75 cm²   5. 3 in.   6. 8 m

7. 7   8. 4   9. 3

Home Link 5-4

1. 350 cm²   2. 60 cm²   3. 279.5 cm²   4. 180 cm²

5. 3.75   6. 27.73   7. 108.59   8. 1

Home Link 5-5

1. c   2. b   4. 15.6

5. 2.08   6. 0.42

Home Link 5-6

2.

Box Design	Surface Area	Number Sentence
Design 1	486 cm²	$6 * 9^2 = 486$
Design 2	507 cm²	$2(5 * 8.7) + 3(10 * 14) = 507$
Design 3	586 cm²	$2(13 * 8) + 2(13 * 9) + 2(8 * 9) = 586$

4. 879.25   5. 373.5   6. 122.67

Home Link 5-7

1. **c.** 70.5 ft²   **d.** 2 pints

2. 64   3. 2.25   4. 1   5. $\frac{4}{9}$

Home Link 5-8

1. Sample answer: Polygon L is a parallelogram. To find its area, multiply the base times the height. The variable *s* is not the height.

3. Sample answer: Divide Polygon K into 2 or more polygons and rearrange the pieces to make a new polygon.

4. 70   5. 360

Home Link 5-9

1. 27 in.³   2. 60 ft³   3. 384 in.³

4. Small: 1,000 cm³; Large: 18,000 cm³; Medium: 8,000 cm³

5. $\frac{9}{6}$ 6. $5\frac{5}{8}$ 7. $\frac{1}{20}$ 8. $9\frac{7}{12}$

Home Link 5-10

1. **Suitcase 1:**   **a.** 2,040 in.³   **b.** 1,430 in.³
   Suitcase 2:   **a.** 2,058 in.³   **b.** 1,457.63 in.³
   Suitcase 3:   **a.** 3,744 in.³   **b.** 2,737.97 in.³
   Suitcase 4:   **a.** 4,788 in.³   **b.** 3,412.5 in.³

3. 4   4. $\frac{5}{7}$ 5. $5\frac{1}{3}$ 6. 3

Home Link 5-11

1. D: 1,600 in.³   O: 1,550 in.³   L: 925 in.³
   I: 1,125 in.³   E: 1,275 in.³

2. 7,400 in.³   3. **a.** $19.99

4. 6.4   5. 405   6. 300   7. 4.4

Home Link 5-12

1. 100 cm   2. 10,000 cm²   3. 1,000,000 cm³

4. 1,000,000 g;   5. 2,200 lb   6. 100 times
   1,000 kg

7. The giant's lungs would provide 100 times as much oxygen, but the giant's body would likely need 1,000 times as much.

8. Sample answer: The lung is like a sponge with many tiny surfaces that total between 70 and 100 square feet of surface area.

9. 7   10. 0   11. −3

Polygon Side Lengths

(1) Find any missing coordinates. Plot and label the points on the coordinate grid. Draw the polygon by connecting the points.

a. Rectangle *ABCD*

A: (1, 1)   B: (−1, 1)

The length of $\overline{BC}$ is represented by

$|1| + |−4| =$ _____.

C: (_____, _____)

D: (_____, _____)

b. Right triangle *XYZ*

X: (−5, 1)   Z: (−3, 6)

The length of $\overline{ZY}$ is represented by $|6| − |1| =$ _____.

The length of $\overline{XY}$ is represented by $|−5| − |−3| =$ _____.

Y: (_____, _____)

(2) Use rectangle *ABCD* and triangle *XYZ* to fill in the following tables. The first row has been done as an example.

Horizontal Sides	Segment Endpoints	Length Expression	Length
$\overline{AB}$	(1, 1) and (−1, 1)	$|−1| + |1|$	2

Vertical Sides	Segment Endpoints	Length Expression	Length

Practice Divide. Write any remainders using R.

(3) $6\overline{)7,329}$ (4) $73\overline{)3,285}$ (5) $38\overline{)8,398}$ (6) $128\overline{)2,310}$

131

Finding the Areas of Parallelograms

Find the area of each parallelogram. Show your work.

(1)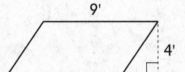

9'

4'

Area: _____

(2)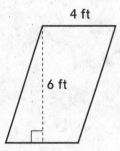

4 ft

6 ft

Area: _____

(3)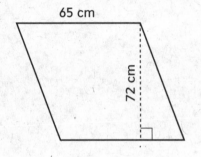

65 cm

72 cm

Area: _____

(4)

8 cm

3.5 cm

Area: _____

Try This The area of each parallelogram is given. Find the length of each base.

(5)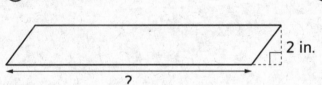

2 in.

?

Area: 26 square inches

Base: _____

(6)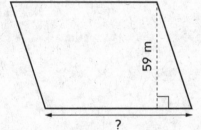

59 m

?

Area: 5,015 square meters

Base: _____

Practice Evaluate.

(7) 20% of 45 _____

(8) 45% of 60 _____

(9) 83% of 110 _____

133

Triangle Area

Find the area of each triangle. Remember: $A = \frac{1}{2}bh$.

(1)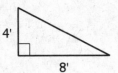

4'

8'

Number model: _____

Area = _____

(2)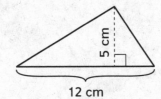

5 cm

12 cm

Number model: _____

Area = _____

(3)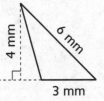

4 mm

6 mm

3 mm

Number model: _____

Area = _____

(4)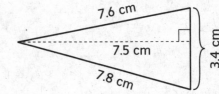

7.6 cm

7.5 cm

3.4 cm

7.8 cm

Number model: _____

Area = _____

(5) Find the length of the base.

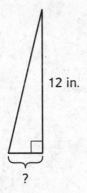

12 in.

?

Area = 18 in.²

Base = _____

Try This

(6) Draw a height for the triangle. Find the length of the height.

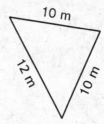

10 m

12 m

10 m

Area = 48 m²

Height = _____

Practice

Compute.

(7) |−7| = ____

(8) |4| = ____

(9) ____ = |−3|

Areas of Complex Shapes

In Problems 1–4, decompose the shapes into polygons for which area formulas can be used. Label the areas. Find the total area for each shape. Use appropriate units.

SRB
257

①

25 cm

15 cm

5 cm

5 cm   15 cm

Area: _____

②

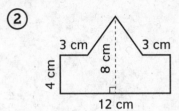

3 cm   3 cm

4 cm   8 cm

12 cm

Area: _____

③

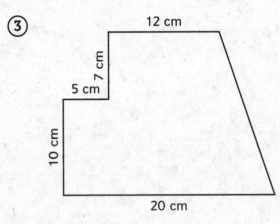

12 cm

7 cm

5 cm

10 cm

20 cm

Area: _____

Try This

④

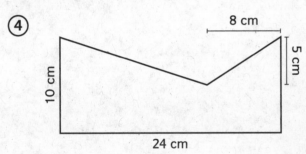

8 cm

5 cm

10 cm

24 cm

Area: _____

Practice Calculate.

⑤ $12 - 8.25 =$ _____

⑥ _____ $= 9.03 + 0.7 + 18$

⑦ $125.29 - 16.7 =$ _____

⑧ _____ $= 0.01 + 0.99$

Real-World Nets

Circle the solid that can be made from each net.

SRB
263-264

①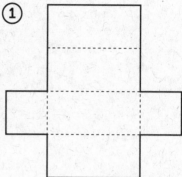

a. 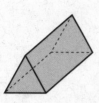   **b.**   **c.**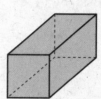

②

a. 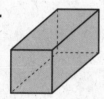   **b.** 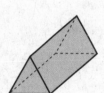   **c.**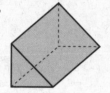

③ Use the net and its corresponding geometric solid in Problem 2.

a. Which polygons make up the faces of your solid?
How many are there of each kind? _____

b. Which faces are parallel? _____

c. Which faces are congruent? _____

d. How many edges are there? How many vertices? _____

Practice Multiply.

④ 5.2 * 3 = _____   ⑤ 1.04 * 2 = _____   ⑥ _____ = 0.14 * 3

Surface Area Using Nets

Silly Socks is trying to choose a type of plastic box for their socks.
The nets for three different box designs are given below.

Design 1

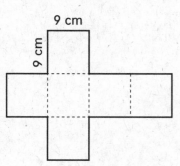

Design 2

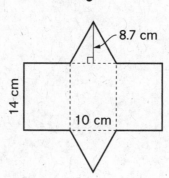

Design 3

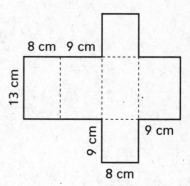

(1) Without calculating, predict which design will require
the least amount of plastic to produce. _____

(2) Find the surface area for each plastic-box design.
Write a number sentence to show how you found the surface area.
Remember to use the correct order of operations.

Box Design	Surface Area	Number Sentence
Design 1		
Design 2		
Design 3		

(3) Explain how to find the surface area for any rectangular or triangular prism.

_____

_____

Practice Divide. Find your answer to the nearest hundredth.

(4) 8)7,034

(5) 18)6,723

(6) 54)6,624

141

Surface Area

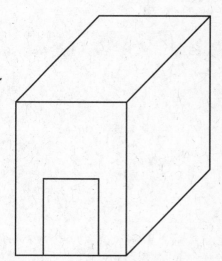

(1) Sam is painting the outside of a doghouse dark green (except for the bottom, which is on the ground).

The doghouse measures 3 feet wide by 4.5 feet long. It is 4 feet high.

The roof is flat, so the doghouse looks like a rectangular prism.

The entrance to the dog house is 1.5 feet wide by 2 feet high.

a. Label the doghouse diagram with the measurements.

b. On the grid below, draw a net for a prism that could represent Sam's doghouse.

Scale: ☐ = 1 square foot

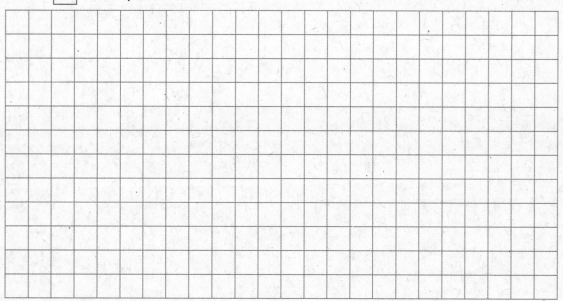

c. How many square feet is he painting? _____

d. One pint of paint covers about 44 ft². How many pints does he need? _____

Practice

Evaluate.

(2) $4^3 =$ _____   (3) $1.5^2 =$ _____   (4) $1^{50} =$ _____   (5) $\left(\frac{2}{3}\right)^2 =$ _____

Arguing about Areas

Jayson was comparing the areas of the polygons at the right.

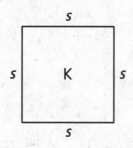
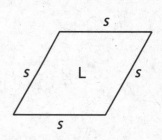

Here is Jayson's reasoning:
I think that Polygons K and L have the same area. I lined up the sides of each polygon and they were equal, so I labeled the sides with the same variables. So the area of Polygon K is equal to the area of Polygon L.

Area of Polygon K:
$s * s = s^2$

Area of Polygon L:
$s * s = s^2$

(1) Explain the flaw in Jayson's reasoning.

_____

_____

_____

Trace Polygon K above, and cut out your tracing. Use it to help you solve Problems 2–3.

(2) Draw two different polygons that have the same area as Polygon K.

(3) Choose one of your polygons from Problem 2. Describe how you used Polygon K to draw a polygon that has the same area.

_____

_____

For Lesson 5-9, bring a rectangular prism, such as an empty tissue box, to class.

Practice Find the whole.

(4) 10% is 7, so 100% is _____.

(5) 25% is 90, so 100% is _____.

145

Volume of Rectangular Prisms

Find the volume for each prism.

SRB
260

①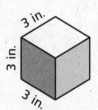

3 in.
3 in.
3 in.

Volume _____

②

5 ft
3 ft
4 ft

Volume _____

③ The Blueberry Blast cereal box is a rectangular prism that is 12 inches × 8 inches × 4 inches.

   a. Label the diagram with the dimensions.

   b. What is its volume? _____

④ Greta's gift shop has three sizes of gift boxes.
They are all shaped like rectangular prisms. The dimensions are shown below.

Small: 10 cm × 10 cm × 10 cm   **Large:** 40 cm × 30 cm × 15 cm

Medium: The area of the base is 1,000 cm² and the height is 8 cm.

Find the volume of each gift box.

Small: _____   **Large:** _____

Medium: _____

Practice   Evaluate.

⑤ $\frac{2}{3} + \frac{5}{6} =$ _____

⑥ $4\frac{3}{4} + \frac{7}{8} =$ _____

⑦ $\frac{4}{5} - \frac{3}{4} =$ _____

⑧ $10 - \frac{5}{12} =$ _____

147

Calculating Luggage Volume

You may want to consider how much volume your luggage holds when you travel. If you know how to calculate the area of a rectangular prism, you can also find the approximate volume of a suitcase. Below are the measurements of some common suitcase sizes.

SRB
260, 262

①   **a.** Find the volume of each suitcase.

       b. Find the approximate volume of the interiors.
          Round to the nearest 0.01 in.^{3}.

Suitcase 1

Exterior: 17" × 15" × 8"

a. Volume: _____

Interior: 16" × 13.75" × 6.5"

b. Volume: _____

Suitcase 2

Exterior: 21" × 14" × 7"

a. Volume: _____

Interior: 19.5" × 13" × 5.75"

b. Volume: _____

Suitcase 3

Exterior: 24" × 16" × 9.75"

a. Volume: _____

Interior: 22.5" × 14.75" × 8.25"

b. Volume: _____

Suitcase 4

Exterior: 28" × 19" × 9"

a. Volume: _____

Interior: 26" × 17.5" × 7.5"

b. Volume: _____

②   Describe how you can estimate the interior volume of a suitcase if you know the exterior measurements.

_____

_____

_____

Practice Evaluate.

③ $\frac{2}{3} \div \frac{1}{6} =$ _____

④ $\frac{5}{12} \div \frac{7}{12} =$ _____

⑤   _____ $= 2\frac{2}{3} \div \frac{1}{2}$

⑥ $8 \div 2\frac{2}{3} =$ _____

Volume of Letters

SRB
260

The Santiago Balloon Emporium sells custom balloons shaped like letters of the alphabet. Clarissa orders balloons that spell DOLLIE for her friend's birthday. She wants the balloons to float, so she plans to fill them with helium. To estimate how much it will cost, Clarissa needs to calculate the approximate volume of helium she will need to fill the balloons.

The volume of each balloon can be estimated based on rectangular prisms.

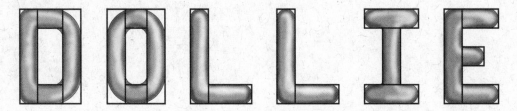

Measure the dimensions in millimeters for each rectangular part of the letters.

(1) The scale is 1 mm = 1 inch. Each letter has a depth of 5 inches. Estimate the volume of each letter.

D: _____   O: _____   L: _____

I: _____   E: _____

(2) What is the approximate total volume of helium (in cubic inches) needed to fill the letters?   _____

(3) **a.** Helium comes in tanks that hold either 8.9 ft³, which cost $19.99 each, or 14.9 ft³, which cost $28.99 each. What is the least amount Clarissa can spend to fill her letters with helium?   _____
Hint: There are 1,728 in.³ in 1 ft³.

b. Explain how you found your answer to Part a.

_____

_____

Practice Divide.

(4) $4\overline{)25.6}$ (5) $0.8\overline{)324}$ (6) $0.16\overline{)48}$ (7) $9.5\overline{)41.8}$

Could a Giant Breathe?

Think about how area and volume change in relation to changes in linear measurements.

SRB
373-378

① How many centimeters are in 1 meter? _____

② How many square centimeters are in 1 square meter? _____

③ How many cubic centimeters are in 1 cubic meter? _____

One cubic centimeter of water has a mass of about 1 gram.

④ One cubic meter of water has a mass of:

_____ grams   _____ kilograms

⑤ One kilogram has a weight equivalent to about 2.2 pounds.
One cubic meter of water weighs about how many pounds? _____

Oxygen enters your body through the **surface area** of your lungs.

⑥ A giant who is 10 times as tall as you would have lungs that provide_____
as much oxygen as your lungs.

⑦ If the surface area of the giant's lungs were 100 times greater than yours,
and if the giant required oxygen in the same proportions as a human,
how do you know the giant would not have enough oxygen? Explain.

_____

_____

Try This

⑧ Your lungs fit in a relatively small space inside your rib cage. Research how your
lungs increase surface area to be able to supply all the oxygen you need.

_____

_____

Practice For Problems 9–10, record the opposite of the number.

⑨ −7 _____   ⑩ 0 _____   ⑪ The opposite of the opposite of −3 _____

153

Unit 6: Family Letter

Equivalent Expressions and Solving Equations

In this unit, your child will explore in depth the algebraic concepts that were introduced in Unit 4. Students will focus on using a variety of models and strategies for solving equations: trial and error, a bar model, a pan-balance model, and an inverse operations strategy. Students start by solving simple equations using trial and error. This helps reinforce the idea, introduced in Unit 4, that solving an equation means finding the value (or values) that can replace a variable to make the number sentence true.

Next students solve equations using a bar model, where the two sides of the equation are represented by the two layers in the bar model. Students then divide the layers into sections such that the corresponding parts line up and students can locate the solution (where a single instance of the variable lines up with a number).

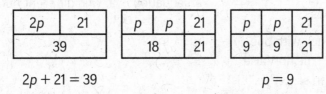

$$2p + 21 = 39 \qquad\qquad p = 9$$

Students then use the pan-balance model for solving equations. For example, a problem might ask how many marbles in the illustration below weigh as much as the cube.

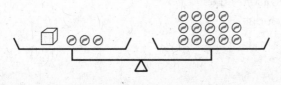

You can find the solution to this example by removing 3 marbles from the left pan and 3 marbles from the right pan. Since you took out the same number of marbles from each pan, the pans will still balance. Therefore the cube weighs the same as 11 marbles.

The pan-balance model helps students transition to writing formal algebraic equations. Students replace the objects in the pans with variables and numbers. Finally, they learn to simplify equations by combining like terms and using inverse operations. For example, $x * (3 + 2) + x = 18$ can be simplified to $5x + x = 18$, and $6x = 18$, and finally $x = 3$. The unit culminates with students generating equations to model situations and to solve related mathematical and real-world problems.

Please keep this Family Letter for reference as your child works through Unit 6.

Vocabulary

Important terms in Unit 6:

constant A quantity that does not change.

equivalent equations Equations that have corresponding equivalent expressions and the same solution set. For example, $2 + x = 4$ and $6 + x = 8$ are equivalent equations. Add 4 to each side of the original to get the new equation. They have a common solution set of {2}. Properly applying the rules of algebra to an equation will produce an equivalent equation.

inverse operations Two operations that reverse each other's effects. For example, addition and subtraction are inverse operations.

open sentence A number sentence with one or more variables that is neither true nor false. For example, $5 + x = 13$ is an open sentence. It is true when $x = 8$.

pan balance A tool used to weigh objects or compare weights. The pan balance is also used as a model in balancing and solving equations.

Pan balance

solution set The set of all solutions of an equation or inequality.

substitute To replace one thing with another, such as a variable with a numeric value.

term In an algebraic expression, a number or a product of a number and one or more variables. For example, in the expression $5y + 3y - 8$, the terms are $5y$, $3y$, and 8. The 8 is a constant term because it has no variable part.

trial and error A systematic method for solving an equation by trying different numbers.

Do-Anytime Activities

1. Ask your child to demonstrate how pan-balance problems work.

2. Encourage your child to generate equations to represent situations at home.
   For example: You and your sister each keep some board games in your rooms, and there are more in the den. You and your sister decide to split up the games in the den evenly so you can store them in your rooms. If x represents the number of board games in your room and y is the total number of board games in the den, how many board games will you have after you and your sister split the games in the den? (Solution: $x + \frac{y}{2}$)

3. Ask your child to make a list of the geometry formulas that he or she has learned so far, such as the one for calculating the area of a rectangle. Have your child identify the variables and constants in the formulas and explain what each variable represents.

Building Skills through Games

In this unit, your child will work on a variety of skills by playing the following games:

Doggone Decimal
Students estimate products of whole numbers and decimals.

Name That Number
Students apply the order of operations in writing expressions equivalent to a target number.

Ratio Comparison
Students compare ratios.

Fraction/Whole Number Top-It
Students multiply whole numbers and fractions and convert improper fractions to mixed numbers.

Algebra Election
Students simplify expressions and substitute numbers for variables to evaluate expressions.

Spoon Scramble (**Expressions**)
Students generate and match equivalent expressions.

Daring Division
Students estimate quotients.

Multiplication Bull's Eye
Students estimate products of 2- and 3-digit numbers.

Landmark Shark
Students find the mean, median, mode, and range of data sets.

As You Help Your Child with Homework

As your child brings assignments home, you might want to go over the instructions together, clarifying them as necessary. The answers listed below will guide you through some of the Home Links in this unit.

Home Link 6-1

1. Sample answer: 3.4, 11.56, 14.96, $<$ 15; 3.45, 11.903, 15.353, $>$ 15; 3.42, 11.696, 15.116, $>$ 15; 3.41, 11.628, 15.038, $>$ 15. Closest number: 3.41

2. Sample answer: 5.7, 32.49, 11.4, 21.09, $<$ 23; 5.8, 33.64, 11.6, 22.04, $<$ 23; 5.9, 34.81, 11.8, 23.01, $>$ 23; 5.89, 34.692, 11.78, 22.912, $<$ 23. Closest number: 5.9

3. 9^3 4. 7^5 5. 6.2^2

Home Link 6-2

1. $7 > j, j < 7$ 2. **a.** $11 < m + 1$

   b. Sample answer: I tried numbers greater than 10 to see which inequalities worked.

3. **a.** {15} **b.** {66}   4. G, D, F, H, B, A, E, C

5. 4.4   6. 0.19   7. 490   8. 990

Home Link 6-3

1. $a = 21$

4a	12
96	

a	a	a	a	12
84				12

a	a	a	a	12
21	21	21	21	12

3. $2s + 6$; $2s + 6 + s = 39$; Jane: 28 years old, Martin: 11 years old

4. $2n + 15 = 5n$, Dan's number: 5

5. 24   6. 14   7. 6.63

Home Link 6-4

1. 3   2. 1   3. 36   4. 2   5. $1\frac{1}{2}$

6. $\frac{5}{3}$ 7. 6   8. $\frac{2}{3}$ 9. $\frac{6}{11}$

Home Link 6-5

Sample answers given.

1. a., b.

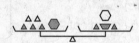 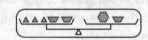

2.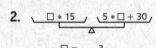

$\square = \underline{\quad 3 \quad}$ $\square = \underline{\quad 24 \quad}$

4. 30.1   **5.** 0.3   **6.** 4.37

Home Link 6-6

1. a. $54x$ **b.** $9w$ **c.** $2.12e$ **d.** $118.5h$

2. a. $36m$ **b.** $60a$ **c.** $29b + 8$ **d.** $36d + 25$

   e. $45f + 121$ **f.** $313 + 128t$

3. $>$ **4.** $>$ **5.** $<$ **6.** $=$

Home Link 6-7

1. a. Sample answer: $12k + 3k + 3 * 2 + 1$

   b. Sample answer: $15k + 7$

2. $3a + 15$ **3.** $10b + 65 + w$

4. $6x + 5x = 8 + 3x - 4, 11x = 3x + 4$

5. $3b - 10 = 10(b + 5) + 15 + b,$
   $b(1 + 2) - 10 = 10b + 50 + 15 + b$

6. Sample model:

$3x + 4$
$11x$

$3x$	4
$3x$	$8x$

$3x$	1	1	1	1
$3x$	$2x$	$2x$	$2x$	$2x$

$x = \frac{1}{2}$

7. 1   **8.** $4\frac{1}{6}$ **9.** $8\frac{13}{40}$ **10.** $25\frac{6}{20}$

Home Link 6-8

1. Katya: $6.25t$; Liova: $2.5t + 100$

2. Sample answer: Katya runs 200 meters in 32 seconds. $6.25t = 200$; $6.25(32) = 200$. Liova runs the last 100 meters in 40 seconds. $2.5t + 100 = 200$; $2.5(40) + 100 = 200$.

3. 3.3   **4.** 9   **5.** 54   **6.** 7,200

Home Link 6-9

1. 5; I added 7 to 8 to get 15, subtracted 5 to get 10, and found half of 10 to get 5.

2. $362 = a$ **3.** $48 = r$

4. $j = 4\frac{1}{4}$ **5.** $1.68 = u$

6. 5.5 pages per minute   **7.** $7.75 per hour

Home Link 6-11

1–4. Models and strategies vary, but $m = 13$, $n = 1$, $p = 1.5$, and $q = 54$.

Approximating Solutions

For each equation, try to get as close as possible to the exact solution. Use the suggested test numbers to get started. Round numbers to the nearest thousandth.

SRB
214-215

① Equation: $r^2 + r = 15$

r	r^2	$r^2 + r$	Compare ($r^2 + r$) to 15
3	9	12	< 15
4	16	20	> 15
3.5	12.25	15.75	> 15

My closest number: _____

② Equation: $x^2 - 2x = 23$

x	x^2	$2x$	$x^2 - 2x$	Compare ($x^2 - 2x$) to 23
6	36	12	24	> 23
5	25	10	15	< 23
5.5	30.25	11	19.25	< 23

My closest number: _____

Practice

Rewrite each expression in exponential notation.

③ 9 * 9 * 9 _____   ④ 7 * 7 * 7 * 7 * 7 _____   ⑤ 6.2 * 6.2 _____

Solution Sets

(1) The solution set is {all numbers less than 7}.
Circle inequalities with this solution set.

$j > 4$ $7 < j$ $7 > j$ $j < 7$

(2) **a.** The solution set is {all numbers greater than 10}.
Circle inequalities with this solution set.

$m + 10 < 11$ $11 < m + 1$ $6 > 5 + m$ $6 > 5m$

b. Explain how you found your answer for Problem 2a.

_____

_____

(3) Record the solution sets for the equations below.

a. $3x = 45$ Solution set: _____

b. $x + 138 = 204$ Solution set: _____

(4) Write the letter of the solution set that matches each number sentence.

$x \div 4 = 8$ _____   **A.** {All numbers}

$\frac{4}{x} = 8$ _____   **B.** {0}

$10 - x = 7$ _____   **C.** { }

$3x + x = 16$ _____   **D.** $\left\{\frac{1}{2}\right\}$

$5x = 0$ _____   **E.** $\left\{-\frac{1}{2}, \frac{1}{2}\right\}$

$12 * x = x * 12$ _____   **F.** {3}

$0.5 = |x|$ _____   **G.** {32}

$x - 5 = x$ _____   **H.** {4}

Practice

Divide.

(5) $8.8 \div 2 =$ _____

(6) $0.95 \div 5 =$ _____

(7) $98 \div 0.2 =$ _____

(8) $198 \div 0.2 =$ _____

161

Modeling and Solving Number Stories

Use bar models to solve these equations. Check your answers.

(1) $4a + 12 = 96$ Solution: _____ Check: _____

(2) $6d + 7 = d + 22$ Solution: _____ Check: _____

Use bar models to solve the problems.

(3) Jane is 6 years older than twice Martin's age.
     Let s be Martin's age.
     Write an expression to represent Jane's age. _____
     The sum of Jane's and Martin's ages is 39.
     Write an equation to represent the situation. _____

     How old are Jane and Martin? Jane: _____ Martin: _____

Try This

(4) Dan is thinking of a number. He doubles his number and adds 15.
     He multiplies his number by 5 and gets the same answer. Let n be Dan's number.

     Write an equation to represent the situation. _____

     Dan's number: _____

Practice   Solve.

(5) $5 * (6.8 - 2) =$ _____ (6) $8 \div 2 * 3.5 =$ _____ (7) $9.43 - 4.5 + 1.7 =$ _____

Pan-Balance Problems

In each figure, the two pans are balanced.
Solve these pan-balance problems.

(1) One triangle weighs

as much as _____ squares.

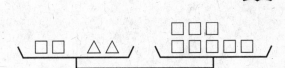

(2) One ball weighs

as much as _____ coin(s).

Note: Remember that 2⬤ means ⬤ ⬤.

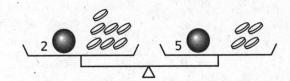

(3) Two cantaloupes weigh

as much as _____ apples.

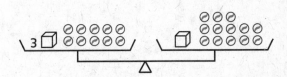

(4) One cube weighs

as much as _____ coin(s).

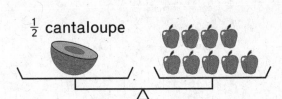

(5) One cube weighs

as much as _____ marbles.

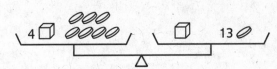

Practice

Solve.

(6) $1 = \frac{3}{5} *$ _____

(7) $\frac{1}{6} *$ _____ $= 1$

(8) _____ $* 1\frac{1}{2} = 1$

(9) $1 = 1\frac{5}{6} *$ _____

165

Solving Pan-Balance Problems

(1) These two pan balances are in perfect balance.

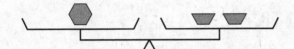 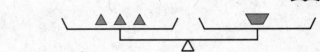

a. Use the relationships in the pan balances shown above to determine which of the pan balances below are balanced. Circle the ones that are in balance.

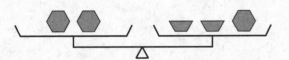 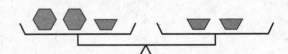

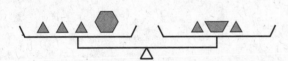 

b. For any pan balance above that you did not circle, add or cross out objects to balance the pans.

(2) Find the value of the missing number that will balance each set of pans below. The same number is missing from both sides of a pan balance.

a. $\Box * 15$ $5 * \Box + 30$

$\Box = $ _____

b. $\Box \div 6$ $\Box - 20$

$\Box = $ _____

(3) Make up two of your own missing-number pan balances.

a.

$\Box = $ _____

b.

$\Box = $ _____

Fill in the missing numbers for the pan-balance problems you made.

Practice Solve.

(4) $4.3 * 7 = $ _____

(5) $0.2 * 1.5 = $ _____

(6) $1.9 * 2.3 = $ _____

167

Simplifying Expressions

(1) Simplify each expression by combining the like terms.

a. $42x + 12x$ _____ **b.** $17w - 8w$ _____

c. $25.42e - 23.3e$ _____ **d.** $88h + 30.5h$ _____

(2) Simplify. Check that your expressions are equivalent.

a. $12m + 24m$ **b.** $90a - 30a$

c. $14b + 15b + 8$ **d.** $58d + 25 - 22d$

e. $3(14 + 15f) + 79$ **f.** $20(18 + 5t) - 47 + 28t$

Practice

Insert =, <, or >.

(3) -5 ____ -10 (4) 0.23 ____ 0.009 (5) -11 ____ 1 (6) 0.092 ____ 0.0920

Exploring Equivalent Equations

1   **a.** Use the Commutative Property (turn-around rule) to create an equivalent expression in which like terms are next to each other.

      $12k + 2 * 3 + 3k + 1$ _____

    b. Combine like terms from Problem 1a
       and write a simplified equivalent expression. _____

Write the expressions in simplest form.

2 $4a + 5 - a + 10$ _____

3 $10(b + 5) + 15 + w$ _____

For Problems 4–5, identify the equations that are equivalent to the given equation.
Circle ALL that apply.

4 $6x + (7 - 2) * x = 8 + 3x - 4$ $6x + 5x = 8 + 3x - 4$

                                                $6x + (7 - 2) * x = 3x + 12$

                                                $11x = 3x + 4$

5 $b + 2b - 10 = 10(b + 5) + 15 + b$ $3b - 10 = 10(b + 5) + 15 + b$

                                                     $b + 2b - 10 = 10b + 50 + 15b$

                                                     $b(1 + 2) - 10 = 10b + 50 + 15 + b$

6   Use a bar model or pan-balance model to solve one of the equations you circled in Problem 4.

      $x =$ _____

Practice

Multiply.

7 $2\frac{1}{3} * \frac{3}{7} =$ _____   **8** $1\frac{2}{3} * 2\frac{1}{2} =$ _____

9   _____ $= 3\frac{7}{10} * 2\frac{1}{4}$ **10** $5\frac{3}{4} * 4\frac{2}{5} =$ _____

Comparing Racing Times

Katya runs at a rate of 6.25 meters per second. Her younger cousin, Liova, runs 2.5 meters per second. Because Katya runs faster than Liova, she gives Liova a 100-meter head start in a 200-meter race.

① Using the variable t to represent the number of seconds, write two expressions—one for Katya and one for Liova—that model how far from the start line they will be after t seconds.

Expression for Katya: _____   Expression for Liova: _____

② Use your expressions from Problem 1 to figure out who will win the race. Show your work and explain your answer.

Winner: _____

Practice

Solve.

③ 5% of 66 is _____.

④ 18% of 50 is _____.

⑤ 45% of 120 is _____.

⑥ 90% of 8,000 is _____.

Using Inverse Operations

(1) Linda has a secret number.

She doubles the number, adds 5, and then subtracts 7. Her result is 8.

What was her original secret number? _____

Explain what you did to find her secret number.

_____

_____

For Problems 2–5, solve the equations using the inverse-operations strategy.
Show all of your steps and check your work.

(2) $257 = a - 105$

(3) $12 = \frac{r}{4}$

Check:

Check:

(4) $j + 3\frac{3}{4} = 8$

(5) $6.72 = 4u$

Check:

Check:

Practice

Write a unit rate for each rate below.

(6) 55 pages in 10 minutes _____

(7) $46.50 for 6 hours _____

175

Solving Pan-Balance Equations

1 Build an equation with two operations that is equivalent to the equation $k = 19$. Record the operations that you use to create each equation below.

Original equation:
Operation (in words)

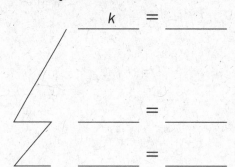

$k \quad = \quad$

_____

_____

$=$

$=$

2 Check that 19 is a solution to your equations.

3 Find the mistake in the work below.

Original pan-balance equation:
Operation (in words)

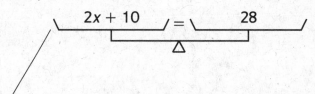

$2x + 10 \quad = \quad 28$

Subtract 10.

$2x \quad = \quad 38$

Divide by 2.

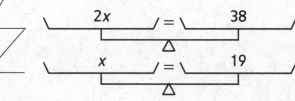

$x \quad = \quad 19$

Describe the mistake and how to correct it.

_____

_____

4 Record the operations you use to create equivalent equations and solve the equation.

Original equation:
Operation (in words)

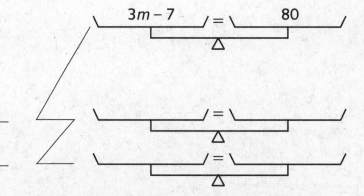

$3m - 7 \quad = \quad 80$

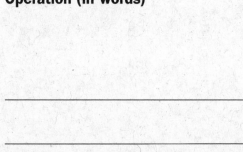

_____

_____

Solving Multistep Equations

Solve the equations below. Use each of these strategies once:

Trial and Error   Bar Model   Pan-Balance Model   Inverse-Operations Strategy

SRB
214-220

Plan ahead to make sure you use the strategy or model that you think works better for each equation.

① $54 = 4m + 2$

② $6n + 8 = 10n + 4$

Strategy: _____

Solution: _____

Strategy: _____

Solution: _____

③ $3p + 3 = 2p + 4.5$

④ $\frac{2}{3}q + 8 = q - 10$

Strategy: _____

Solution: _____

Strategy: _____

Solution: _____

Practice   List the numbers in order from least to greatest.

⑤ $\frac{9}{4}$, 2.35, $\frac{1}{8}$, 1.5, $\frac{3}{8}$

Unit 7: Family Letter

Variables and Algebraic Relationships

Unit 7 builds on the algebraic concepts students learned in Units 4 and 6. To begin, students practice representing real-world situations with inequalities and solving the inequalities to find solution sets for the real-world situations.

This unit introduces students to computer spreadsheets. Students write formulas to perform calculations based on spreadsheet cells and then create graphs from the spreadsheet entries. If you use computer spreadsheets at work or at home, you might want to share your experiences with your child.

Students use proportional reasoning, measurement conversions, and unit rates to compare quantities in real-world situations. For example, they compare the sugar content of various drinks using information from nutritional labels (as shown below) to determine which drink has the greatest concentration of sugar.

Nutrition Facts
Serving Size 1 cup (8 fl oz)

Amount Per Serving	
Calories 112 Calories from Fat 0	
	% Daily Value*
Total Fat 0 g	0%
Saturated Fat 0 g	0%
Trans Fat 0 g	
Cholesterol 0 mg	0%
Sodium 2 mg	0%
Total Carbohydrate 32 g	8%
Dietary Fiber 0 g	0%
Sugars 32 g	
Protein 0 g	

Vitamin A 0%		Vitamin C 96%	
Calcium 0%		Iron 0%	

* Percent Daily Values are based on a 2,000 calorie diet. Your daily values may be higher or lower depending on your calorie needs.

Nutrition Facts
Serving Size 1 can (12 fl oz)
Serving Per Container 1

Amount Per Serving	
Calories 140	
	% Daily Value*
Total Fat 0 g	0%
Saturated Fat 0 g	0%
Trans Fat 0 g	
Cholesterol 0 mg	0%
Sodium 45 mg	2%
Total Carbohydrate 39 g	13%
Dietary Fiber 0 g	0%
Sugars 32 g	
Protein 0 g	0%

* Percent Daily Values are based on a 2,000 calorie diet. Your daily values may be higher or lower depending on your calorie needs.

In the second half of the unit students continue their work with equations, making connections between equations and representations of situations in the form of tables and graphs. They explore relationships between dependent and independent variables and interpret graphical representations of those relationships. Unit 7 starts to formalize the algebra explored so far in sixth grade.

Please keep this Family Letter for reference as your child works through Unit 7.

Vocabulary

Important terms in Unit 7:

dependent (responding) variable (1) A *variable,* whose value is dependent on the value of at least one other variable in a function. (2) The output of a "What's My Rule?" table.

formula A general rule for finding the value of something. A formula is often written using letters, called variables, which stand for the quantities involved. For example, the formula for the area of a rectangle may be written as $A = l * w$, where A represents the area of the rectangle, l represents its length, and w represents its width.

independent (manipulated) variable (1) A *variable* whose value does not rely on the values of other variables. (2) The input of a "What's My Rule?" table.

spreadsheet program A computer application in which information is arranged in cells in a grid. The computer can use the information in the grid to perform operations and evaluate formulas. When the value in a cell changes, the values in all other cells that depend on it are automatically changed.

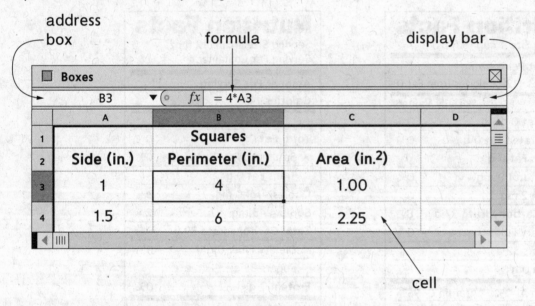

Do-Anytime Activities

1. Ask your child to show you how to represent problem situations using equations. For example, if a babysitter makes $10.00 an hour, you can write the total amount earned (S) as $10 * p = S$, where p stands for the number of hours. You can use this equation to find out what the sitter makes for any given number of hours.

2. While grocery shopping, have your child compare nutrition labels of various foods to determine the sugar, protein, carbohydrate, or fat concentration.

3. If you use a spreadsheet program on your computer, help your child learn how to use it. For example, you might ask your child to set up a sheet to track his or her math scores, graph the data, and find the mean.

Building Skills through Games

In Unit 7, your child will work on a variety of skills by playing the following games:

Hidden Treasure
Students use coordinates and coordinate grids.

Spoon Scramble
Students find fraction, decimal, and percent parts of the whole.

Multiplication Top-It
Students practice their multiplication skills.

Getting to One
Students compare decimals and use patterns and proportional reasoning to determine a mystery number in the fewest tries.

Ratio Comparison
Students compare ratios, using ratio/rate tables to match given ratio situations.

Solution Search
Students solve open sentences.

As your child brings assignments home, you might want to go over the instructions together, clarifying as necessary. The answers listed below will guide you through this unit's Home Links.

Home Link 7-1

1. **a.** Sample answers: $f - 5 > 7$; $f < 13$
   Sample answers: {All numbers greater than 12}; {All numbers less than 13}

2. Sample answers:

   a. Inequality A: $d < 7$
   Inequality B: $d > 3$

   b. Inequality C: $d + 4 < 11$
   Inequality D: $d - 0.5 > 2.5$

3. 4   **4.** 0.5   **5.** 6 or −6

Home Link 7-2

1. **a.** $5y \leq 9$ **b.** {All numbers less than or equal to 1.8 and greater than 0}   **c.** $y \leq 1.8$ and $y > 0$

2. **a.** $0.75m \leq 3$ **b.** {All whole numbers less than or equal to 4} (or {0, 1, 2, 3, 4})   **c.** $y \leq 4$ and $y \geq 0$

3. Sample answer: You can have 0, 1, 2, 3, or 4 ingredients (not 2.5).

4. $\frac{3}{2}$ **5.** $\frac{1}{5}$ **6.** $\frac{4}{15}$

Home Link 7-3

1.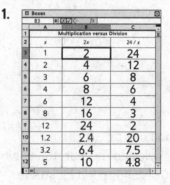

2.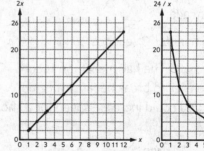

3. One is a line that increases when x does; the other is a curve that decreases when x does.

4. 2   **5.** 7   **6.** 15

Home Link 7-4

1. **a.** 28 pennies, 48 nickels, 24 dimes

   b. $5.08

2. Sample answer: Because you will be dividing by 2 for the number of dimes, which only come in whole quantities

3. $= 0.01 * A2 + 0.05 * B2 + 0.1 * C2$

4. 248, 249, 250, 251

5. 7   6. $\frac{18}{7}$ 7. $\frac{5}{9}$

Home Link 7-5

2. Watching TV

3. Sample answer: He burns about 357 calories sleeping and about 504 calories at school. He burns more at school.

4. Soccer; I divided the calories by the number of minutes. Running burns 9.3 calories/minute; biking burns 8.3 calories/minute; and soccer burns 9.5 calories/minute.

5. 48   6. 42   7. 200

Home Link 7-6

1. **a.** Half marathon

   b. The half-marathon unit rate is about 310 seconds per mile and the marathon unit rate is about 320 seconds per mile, so she ran about 10 seconds faster per mile for the shorter race.

2. 2 hours 15 minutes 8 seconds

3. Sample answer: The half marathon record would be easier to break because it is more likely that you could increase speed for a short distance.

4. 9   5. 166   6. 101

Home Link 7-7

1. **a.** 22 gal   **b.** 154 gal

2. **a.** 7 gal   **b.** 2,555 gal

3. Sample answer: The dishwasher, because it takes about 16 gal per week and hand washing takes about 105 gal.

4. Sample answer: Yes. A dishwasher saves enough to fill the pool (about 138 gal per week or 7,176 gal per year).

5. True   6. True

Home Link 7-8

2. Sample answer: $x * 3 + 2 = y$

4. 12   5. 90   6. 1,870

Home Link 7-9

1. $220 - x = y$

3. Age is independent because age determines the heart rate.

5. −4   6. 9   7. 1.5

Home Link 7-10

3. Sample answer: The area grows faster than the perimeter. Area change increases. Perimeter change remains constant.

4. 40   5. 60

Home Link 7-11

2. 144   3. 201   4. 265

Mystery Numbers

SRB
210-211

(1) Gabe and Aurelia play *Number Squeeze*.
Gabe represents his mystery number with the variable *f*.

  a. Represent each of the two *Number Squeeze* clues with an inequality.
Describe the solution sets to the inequalities.

Clue	Subtract 5 from *f* and the answer is greater than 7.	The number *f* is less than 13.
Inequality		
Solution Set		

  b. Graph the solution set that makes both inequalities true.

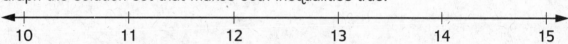

     10   11   12   13   14   15

  c. List three numbers that could be the mystery number.
Check that they are in the solution sets for both inequalities.

     Possible numbers *f* could be: _____

(2) **a.** Write two inequalities that could be clues for the following graph:

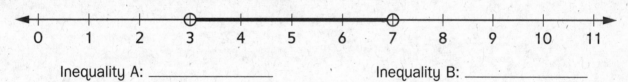

     0   1   2   3   4   5   6   7   8   9   10   11

     Inequality A: _____   Inequality B: _____

  b. Write a different set of inequalities that could also represent the graph in Problem 2a.

     Inequality C: _____   Inequality D: _____

Practice   Evaluate.

(3) $|-4| =$ _____   (4) $|-0.5| =$ _____   (5) $|z| = 6; z =$ _____

Solving Problems with Inequalities

Fast and Healthy sells bags of trail mix. Customers choose the ingredients to put in their trail mix. The bag is weighed at the checkout counter to determine the cost. Fast and Healthy charges $5 per pound. They also sell granola bars for $1.50 each.

(1) Li has $9 to spend on trail mix. How many pounds of trail mix can she buy? Let y be the number of pounds of trail mix.

   a. Inequality for the situation: _____

   b. Solution set for y using set notation: _____

   _____

   c. Inequalities for the values of y: _____

   d. Graph the solution set for y that makes both inequalities true.

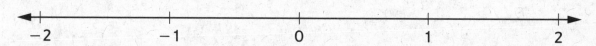

(2) The price for a plain smoothie is $2.00. Each additional ingredient costs $0.75. Li has $5. Let m be the number of ingredients. How many ingredients can Li add to a plain smoothie?

   a. Inequality for the situation: _____

   b. Solution set for m using set notation: _____

   _____

   c. Inequalities for the whole number values of m: _____

   d. Graph the solution set for m that makes both inequalities true.

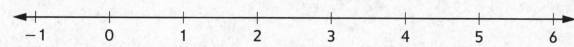

(3) Describe how the graph in Problem 2d represents the solution to the problem.

_____

_____

Practice

Solve.

(4) $\frac{2}{3}$ * _____ = 1   **(5)** _____ * 5 = 1   **(6)** $3\frac{3}{4}$ * _____ = 1

Using Spreadsheets

① Complete the spreadsheet at the right. If you have a spreadsheet program at home, write formulas and use the "fill down" feature to do the calculations. If not, do the calculations yourself with a calculator.

② Use the data in the spreadsheet to graph the number pairs for x and 2x on the first grid. Then graph the number pairs for x and 24 / x on the second grid. Connect the plotted points.

Boxes			⊠
B3	◆ X √ ⊙	fx	
	A	B	C
1	**Multiplication versus Division**		
2	x	2x	24 / x
3	1		
4	2		
5	3		
6	4		
7	6		
8	8		
9	12		
10	1.2		
11	3.2		
12	5		

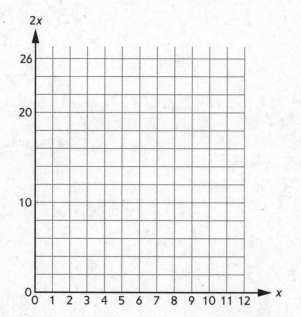

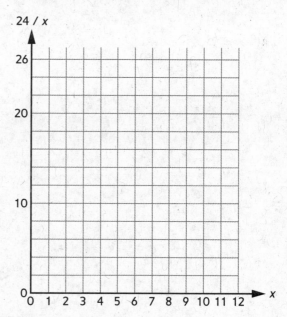

③ Describe two differences between the two graphs in Problem 2.

_____

_____

Practice Find the GCF.

④ GCF (34, 42) = _____   ⑤ GCF (49, 560) = _____   ⑥ GCF (30, 75) = _____

189

Using a Spreadsheet

Use a spreadsheet program or your calculator to complete the page.

1. Jenna has a large jar full of pennies, nickels, and dimes. She has 100 coins.
   She has 20 more nickels than pennies and half as many dimes as nickels.
   Enter formulas in the spreadsheet to calculate the number of coins and their value.

	A	B	C	D
1	Pennies	Nickels	Dimes	Total
2	Test Number			

   a. Coins: _____

   b. Value of coins: _____

2. To solve Problem 1a, why might you start with an even number of pennies?

   _____

   _____

3. What formula would you use to find the total value of the coins?

   _____

4. Use formulas to find the greatest four consecutive
   numbers whose sum is less than 1,000. _____

	A	B	C	D	E
1	1st Number	2nd Number	3rd Number	4th Number	Sum
2	Test Number				

Practice Divide.

5. $\frac{7}{8} \div \frac{1}{8} =$ _____

6. $2\frac{1}{4} \div \frac{7}{8} =$ _____

7. _____ $= 1\frac{2}{3} \div 3$

191

Which Activity Burns the Most Calories?

(1) The amount of energy a food will produce when it is digested by the body is measured in a unit called the **calorie.**

The table shows the number of calories used per minute and per hour by the average sixth grader in Oakwood Junior High for various everyday activities. Complete the table. Use the information for Problems 2–3.

Calorie Use by the Average Sixth Grader		
Activity	Calories/Minute	Calories/Hour
Sleeping	0.7	42
Studying, Writing, Sitting	1.2	
Standing	1.3	
Dressing, Undressing		90
Watching TV	1.0	
Eating, Talking		72

(2) Kori spent 2 hours and 25 minutes doing one of the listed activities.

He burned 145 calories. Which activity was he doing? _____

(3) Kori sleeps about $8\frac{1}{2}$ hours per night and spends about 7 hours each school day eating, talking, and sitting. Does he burn more calories sleeping or at school? Explain.

_____

_____

(4) On Monday Edgar ran for 29 minutes and burned 270 calories.
On Wednesday he biked for 25 minutes and burned 207 calories.
On Friday he played soccer for 13 minutes and burned 124 calories.
Which activity burns the most calories per minute? Explain how you know.

_____

_____

_____

Practice Find the LCM.

(5) LCM (12, 48) = _____   (6) LCM (14, 21) = _____   (7) LCM (8, 25) = _____

Marathon Mathematics

In 2006, Deena Kastpor set the U.S. women's record for both the half marathon (13.1 miles) and the full marathon (26.2 miles).
Her time for the half marathon was 1 hour 7 minutes 34 seconds.
Her time for the full marathon was 2 hours 19 minutes 36 seconds.

(1) Compare her rates (seconds per mile) for the two races.

    a. Which rate was faster? _____

    b. How much faster is her rate for that race?

    _____

    _____

    _____

    _____

(2) If Deena could run a full marathon at her half-marathon pace, about how long would it take her to run the full marathon?

    _____

(3) Which record do you think would be easier to break: half marathon or full marathon? Explain.

    _____

    _____

    _____

Practice Find the value of x that makes each number sentence true.

(4) $6x = 54$ _____   **(5)** $x - 14 = 152$ _____   **(6)** $300 = x + 199$ _____

Doing the Dishes

① Ronald's family washes dishes by hand.
Hand washing the dinner dishes takes about 10 minutes,
and the faucet is running the whole time.
The kitchen faucet runs at about 2.2 gallons per minute.

   a. In one evening, about how much water does
      Ronald's family use to wash dinner dishes?   _____

   b. In seven evenings, about how much water
      does the family use to wash dishes?   _____

② A high-efficiency faucet runs at about 1.5 gallons per minute.

   a. About how much water would the family save each time
      they wash their dinner dishes if they replace their old
      faucet with a high-efficiency faucet?   _____

   b. About how much water would they save
      washing dinner dishes in a year (365 days)?   _____

③ A high-efficiency dishwasher uses about 4 gallons of water per load. The family would
run the dishwasher 4 times per week to do their dinner dishes. Should they install a
high-efficiency faucet (see Problem 2) or use the dishwasher to save water? Explain.

_____

_____

Try This

④ A typical circular pool that is 18 feet across and 4 feet deep requires about
3,800 gallons of water. Ronald's parents agree to get this pool if they cut their water
usage enough to fill the pool. If they use the dishwasher, can Ronald's family save
enough water during the year to justify getting the pool? Explain.

_____

_____

Practice   Write whether each number sentence is true or false.

⑤ $4 * 7 > 6 * 3 + 4$ _____   ⑥ $15 + 9 \leq 6 * 4$ _____

SRB
43-50

Representing Patterns in Different Ways

Use the pattern below to answer the questions.
Hint: The perimeter of one trapezoid is 5, not 4.

SRB
225

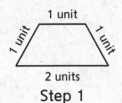
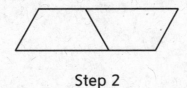
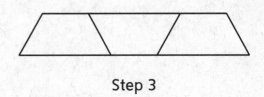

Step 1 Step 2 Step 3

1 unit
1 unit
1 unit
2 units

① In the space below, sketch and label Step 4 and Step 5 of the sequence.

② Complete the table, and record an equation to represent the rule for finding the perimeter.

Rule: _____

Step Number (x)	Perimeter (y) (units)
1	
2	
3	
4	
5	
10	

③ Use the values in the table from Problem 2 as the *x*- and *y*-coordinates for points. Graph the points on the coordinate grid.

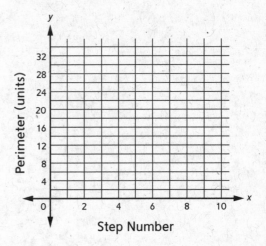

Perimeter (units)

Step Number

Practice

Evaluate.

④ 15% of 80 = _____ ⑤ 45% of 200 = _____ ⑥ 85% of 2,200 = _____

199

Maximum Heart Rate

One way you can tell whether you are exercising too much, too little, or just the right amount is to check your heart rate. Calculate the number of beats per minute. The ideal average maximum heart rate is calculated by subtracting your age from 220.

1. Write an equation that represents the rule for calculating your ideal maximum heart rate. Rule: _____

2. Use your rule to complete the table at the right with the beats per minute.

Age (x)	Max. Heart Rate (y)
5	
12	
20	
45	
60	

3. Explain how you know which variable is independent and which is dependent.

_____

_____

4. Graph the values in the table from Problem 2 as the x- and y-coordinates for points.

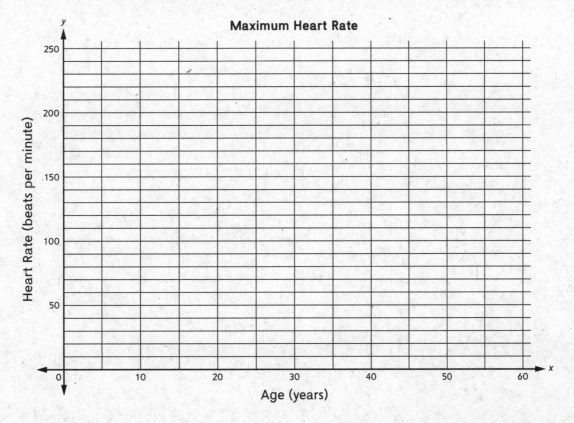

Maximum Heart Rate

(y-axis: Heart Rate (beats per minute), x-axis: Age (years))

Practice Evaluate.

5. −(4) = _____

6. −(−9) = _____

7. −(−1.5) = _____

Comparing Tables and Graphs

① Complete the tables for squares with the given side lengths.

Side Length (in.)	Perimeter (in.)
1	
2	
3	
4	
5	

Side Length (in.)	Area (in.²)
1	
2	
3	
4	
5	

② Use the values in the tables in Problem 1 to make graphs for perimeter and area.

Perimeter

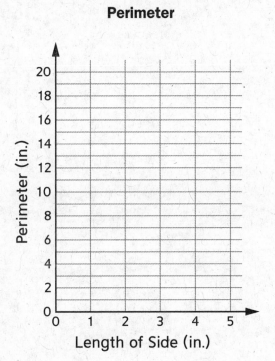

Area

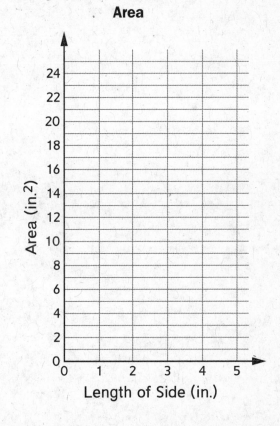

③ Explain why the graphs look different.

_____

_____

Practice Find each number based on the given percents.

④ 10% of *n* is 4; *n* = _____

⑤ 30% of *n* is 18; *n* = _____

Mystery Graphs

① Create a mystery graph on the grid below. Be sure to label the horizontal and vertical axes. Describe the situation that goes with your graph on the lines provided.

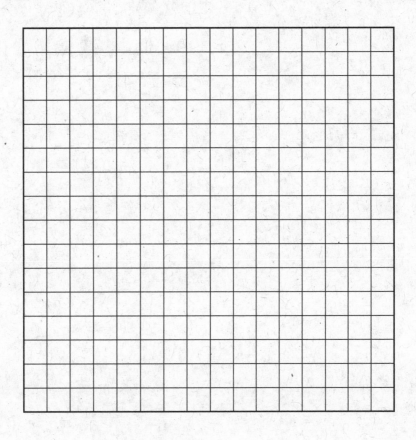

_____

_____

_____

_____

_____

_____

Practice Compute. Use the back of the page to do the computation.

② $3\overline{)432}$ ③ $12\overline{)2,412}$ ④ $5\overline{)1,325}$

Applications: Ratios, Expressions, and Equations

In Unit 8, students apply the mathematical concepts and skills they learned throughout sixth grade to answer real-world questions. Students explore the following situations in this unit:

- *Applying unit rates and proportional reasoning*—Students explore the advantages of square-foot gardening versus row gardening.

- *Interpreting scale with ratios*—Students use a scale factor to calculate and compare scale drawings of artwork to the actual size of artwork on a gallery wall.

- *Scaling up distances on a coordinate grid*—Students devise a plan to hang artwork of various dimensions on a gallery wall.

- *Using ratios to make a model*—Students create a scale-model of the solar system.

- *Comparing ratios*—Students calculate population densities for different parts of the world.

- *Using equivalent expressions and equations*—Students design and make mobiles and balance them using equations.

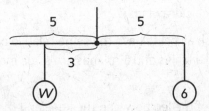

- *Working with dependent and independent variables*—Students use tables, formulas, and graphs to investigate the relationship between the measurements of different human body components.

- *Using variables and equations*—Students use spreadsheets and formulas to create a budget for a family road trip.

Please keep this Family Letter for reference as your child works through Unit 8.

Vocabulary

Important terms in Unit 8: The unit vocabulary reflects the words used in the real-world exploration situations.

anthropometry The study of human body sizes and proportions. An **anthropometrist** is a person who gathers data on the size of the human body and its components.

circumference The distance around a circle.

dwarf planet A celestial body that orbits the Sun and has a spherical shape but is not large enough to disturb other objects from their orbits.

enlarge To make something larger.

fulcrum (1) The point or place around which a lever pivots. (2) A point on a mobile at which a rod is suspended.

mobile A sculpture constructed of rods and other objects that are suspended in midair by wire, twine, or thread.

population density The ratio of people to area of a region calculated by dividing the number of people by the area of the region.

prediction line A line representing all the points for which a formula is true.

scale drawing A drawing of an object or a region in which all the parts are drawn to the same scale. Architects and builders use scale drawings.

scale model A model of an object in which all parts have the same relative proportions as the actual object. For example, many model trains and airplanes are scale models of actual vehicles.

tibia The bone that forms the front part of the leg between the knee and the ankle.

yield The amount produced (as in a garden).

Do-Anytime Activities

1. Ask your child to show you how to choose a scale for making a scale model of your house.

2. Research the ideal planting conditions for different vegetables, and have your child use the information to design square-foot garden plots for a selection of plants.

3. Have your child use spreadsheets to develop budgets for various family events, such as a graduation party, family picnic, or road-trip vacation. Ask your child to explain how to use formulas in the spreadsheet to calculate totals.

Building Skills through Games

In Unit 8, your child will work on a variety of skills by playing the following games:

First to 100
Students use algebraic equations to answer questions.

Algebra Election
Students solve equations.

Ratio Comparison
Students compare ratios and use ratio tables to match given ratio situations.

Solution Search
Students solve open sentences.

As your child brings assignments home, you might want to go over the instructions together, clarifying as necessary. The answers listed below will guide you through this unit's Home Links.

Home Link 8-1

1.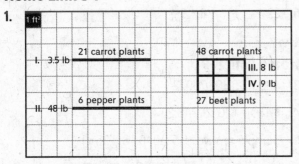

2. **a.** 68.5 pounds   **b.** About $\frac{2}{3}$ plant to 1 ft²

3. **a.** 2 more pounds

   b. Just over 1 plant to 1 ft²

4. 174   5. 84   6. 96

Home Link 8-2

1. **a.** $\frac{1}{2}$ inch = 1 foot
   b. Scaled Dimensions: 2" × 1.5"; 1.25" × 1";
   1.25" × 0.75"; 1" × 1.25"; 1.5" × 1"; 1.25" × 1.5"
   Actual Dimensions: 4' × 3'; 2.5' × 2';
   2.5' × 1.5'; 2' × 2.5'; 3' × 2'; 2.5' × 3'

2. Sample answer: Every $\frac{1}{4}$ inch equals 6 inches; 2 squares equals a foot. Four squares wide is 2 feet; 5 squares high is 2.5 feet.

3. Sample answer: To make sure they like how it looks before they put nail holes in the wall

4. 40   5. 20

Home Link 8-3

1.

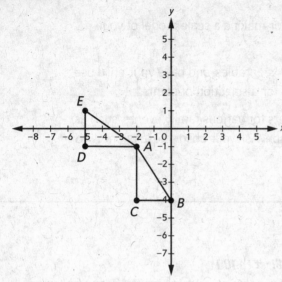

2. (2, −7); (−2, −7); (−8, −1); (−8, 3)

3. 4 units **4.** 2 units **5.** 2 : 1

6. $2\frac{1}{2}$ **7.** $3\frac{1}{6}$ **8.** 3

Home Link 8-4

1. 58 cm; 110 cm; 150 cm; 230 cm; 780 cm; 1,400 cm; 2,900 cm; 4,500 cm; 5,900 cm

2. Sample answer: No. The scale is too large for the distances. The model would not fit in the classroom.

3. Sample answer: 1 cm = 10 million kilometers

4. $t = 96$ **5.** $p = 99$ **6.** $n = 4.65$

Home Link 8-5

1. 57,600 ft²; 2,618 ft² **2.** 4,700 ft²; 470 ft²

3. 17,000 ft²; 1,417 ft² **4.** 108,500 ft²; 10,850 ft²

5. a. Basketball

  b. Sample answer: Basketball has the smallest area but about the same number of players as other sports.

6. Sample answer: The larger the number of square feet per player, the more spread out the players are.

7. Sample answer: Each player is responsible for covering more of the playing field.

8. 3t **9.** 5r + 3.5 **10.** 27c **11.** 2b + 6

Home Link 8-6

1. $W = 3x$; $D = 8$; $w = 15$; $d = 8$; $3x * 8 = 15 * 8$; $x = 5$; 15

2. $W = 8$; $D = x + 4$; $w = 16$; $d = x − 4$; $8(x + 4) = 16(x − 4)$; $x = 12$; 16 lb; 8 lb

4. 69 **5.** 1.8 **6.** 0.305 **7.** 56

Home Link 8-7

1. a. $5 + 2 * 0$; $5 + 2 * 1$; $5 + 2 * 2$; $5 + 2 * 3$
  b. Sample answer: $5 + 2 * (n − 1)$

2. Sample answer: $3 + 2n + n + 1$

3. Sample answer: The expression $3 + 2n + n + 1$ is the best match because it shows the row of three shaded, the multiple of 2 added on top, the number added at the bottom, and the one shaded below.

4. 5x + 50 **5.** 2x + 16 **6.** 3x

Home Link 8-8

3. Sample answers: $w = 2t$; $n = 2w$; $s = 2n$

5. 22 **6.** $3\frac{5}{9}$

Home Link 8-9

3. $300; $5,368,709.12

4. $4,650; $10,737,418.23

5. Sample answers: 5 : 4, 10 : 8, 1.25 : 1

6. Sample answers: 2 : 2.8, 20 : 28, 40 : 56

7. Sample answers: 1 : 6, 2 : 12, 5 : 30

8. Sample answers: 1 : 1.5, 2 : 3, 4 : 6

Increasing Yield per Square Foot

The diagram below shows the layout of a garden that has both rows and squares.

Use the yields on the diagram and the table information to determine which plant is in each row or square foot.

Plant	Distance between Plants	Plant Yield Rate
Beets (1 beet per plant)	4"	3 beets per lb
Carrots (1 carrot per plant)	3"	6 carrots per lb
Lettuce	6"	2 lb per plant
Peppers	12"	8 lb per plant

① Label each garden bed to show what kind of plant and how many of the plants fill the row or square foot.

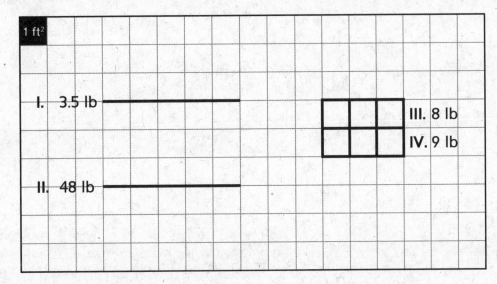

I. 3.5 lb

II. 48 lb

III. 8 lb

IV. 9 lb

1 ft²

② **a.** What is the total expected yield for the garden in Problem 1? _____

b. What is the overall rate of plants per square foot? _____

③ **a.** About how much more should the garden yield if beds I and II are changed from row garden beds to square-foot garden beds?
(Assume the same plant would still be planted in each.) _____

b. What would the overall rate of plants per square foot be? _____

Practice

Solve the equations.

④ $\frac{1}{2}p = 87$; $p =$ _____

⑤ $\frac{2}{3}d = 56$; $d =$ _____

⑥ $\frac{7}{8}k = 84$; $k =$ _____

211

Using Scale Drawings

1 Julian made a scale drawing of his bedroom wall. He has artwork that he and his brother made hanging on the wall. His wall is 7 feet high and 12 feet long.

Letter	Scaled Dimensions (height × width)	Actual Dimensions (height × width)
A		
B		
C		
D		
E		
F		

a. What scale did he use?

_____

b. Use his scale drawing to complete the table.

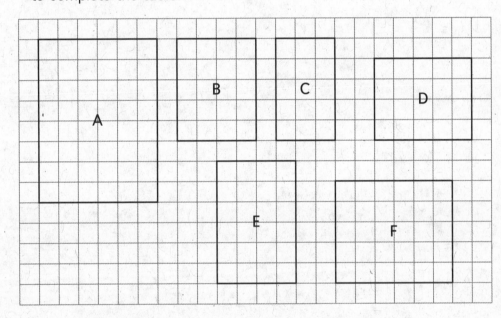

2 Explain how you found the actual dimensions for Artwork B.

_____

_____

3 Why might someone make a scale drawing of a planned artwork arrangement?

_____

_____

Practice Solve.

4 15% of x is 6. 100% of x is _____.

5 90% of y is 18. 100% of y is _____.

213

Stretching Triangles

1 Plot the original ordered pairs from the table in Problem 2, and connect points to make triangle *ABC* and triangle *ADE*.

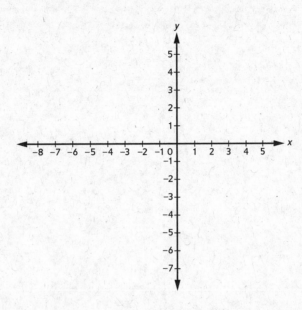

2 If you want to make triangle *ABC* and triangle *ADE* twice as tall and twice as wide, what would the new coordinates be? Write them in the table below.

Point	Original Ordered Pair	Ordered Pair for Enlargement
A	(−2, −1)	(−2, −1)
B	(0, −4)	
C	(−2, −4)	
D	(−5, −1)	
E	(−5, 1)	

3 What is the distance from *D* to *E* in the enlarged figure? _____

4 What is the distance from *D* to *E* in the original figure? _____

5 Use ratio notation to represent the ratio of side length $\overline{DE}$ in the enlarged figure to side length $\overline{DE}$ in the original figure. _____

Practice   Solve.

6 $\frac{3}{4}\left(4 - \frac{2}{3}\right) =$ _____

7 $3 + \frac{1}{2} \div 3 =$ _____

8 _____ $= 2\frac{1}{2} \div \left(\frac{4}{3} - \frac{1}{2}\right)$

Modeling Distances in the Solar System

Today the class made scale models of celestial bodies.
Imagine you are modeling the distance of each planet from the Sun.

① Calculate the distance from the Sun in your model using the scale given in the table.
Complete the table.

Celestial Body	Average Distance from the Sun (km)	Average Distance from the Sun for the Model (Scale: 1 cm = 1,000,000 km)
Mercury	58,000,000	
Venus	110,000,000	
Earth	150,000,000	
Mars	230,000,000	
Jupiter	780,000,000	
Saturn	1,400,000,000	
Uranus	2,900,000,000	
Neptune	4,500,000,000	
Pluto	5,900,000,000	
Sun		

② Would this scale work for building the model in your classroom? Why or why not?

_____

_____

③ What scale for distance might work for a model in your classroom?

_____

Practice

Solve.

④ $\frac{t}{12} = 8$ _____

⑤ $p \div 9 = 11$ _____

⑥ $n + 0.35 = 5$ _____

Comparing Player Density

The dimensions of the playing surfaces for four sports are listed below.

Football: 360 ft by 160 ft   **Hockey:** 200 ft by 85 ft (Ignore round corners)

Basketball: 50 ft by 94 ft   **Baseball:** 108,500 ft² (Average for major league parks)

SRB
49-50

During a game, there are 22 players on a football field, 10 on a basketball court, 10 on a baseball diamond (not counting base runners), and 12 on an ice hockey rink. Calculate the square feet of playing area per player for each sport.

① Football playing area: _____   Area per player: _____

② Basketball playing area: _____   Area per player: _____

③ Hockey playing area: _____   Area per player: _____

④ Baseball playing area: _____   Area per player: _____

⑤ **a.** Which sport is the most "crowded"? _____

   b. Justify your answer.

   _____

   _____

⑥ Describe the relationship between square feet per player and player density.

   _____

   _____

⑦ If the player density is lower, how might that affect their role in the game?

   _____

   _____

Practice Simplify the expressions.

⑧ $7t - 4t$ _____   ⑨ $5 + 7r - 1.5 - 2r$ _____

⑩ $9(3c)$ _____   ⑪ $\frac{1}{2}(4b + 12)$ _____

Mobiles

The mobiles shown in Problems 1 and 2 are in balance.
All measures are in feet for distances or pounds for weight.

SRB
226-227

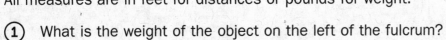

(1) What is the weight of the object on the left of the fulcrum?

$W =$ _____ $D =$ _____

$w =$ _____ $d =$ _____

Equation: _____

Solution: _____   Weight: _____

(2) What is the distance of each object from the fulcrum?

$W =$ _____ $D =$ _____

$w =$ _____ $d =$ _____

Equation: _____

Solution: _____

Distance on the left of the fulcrum: _____

Distance on the right of the fulcrum: _____

(3) **a.** Sketch a mobile that will balance.
Label all lengths and weights.

b. Use the mobile formula to explain why your mobile balances.

_____

_____

Practice

Divide.

(4) $34.5 \div 0.5 =$ _____   (5) $8.46 \div 4.7 =$ _____

(6) _____ $= 1.22 \div 4$ (7) _____ $= 26.88 \div 0.48$

221

Generalizing Patterns

①   Use the pattern pictured below. Shade a constant part of the pattern.

    SRB 225

   a. In the table, record numeric expressions
for the total number of squares that represent
what is shaded and how the pattern is growing.

      Plan your numeric expressions to show
one part of the expression as constant
and the other part as varying.

   b. Write an algebraic expression
for the number of squares in Step n.

Step Number	Expression for Number of Squares
1	
2	
3	
4	

_____

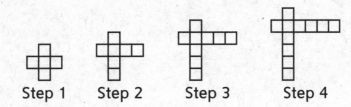

     Step 1   Step 2   Step 3   Step 4

②   Circle an expression for finding the number of squares needed to build Step n that
best represents how the pattern is shaded.

$3 + 2n + n + 1$

$4 + 3n$

$6 + n + 2(n - 1)$

$7 + 3(n - 1)$

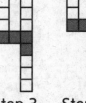

Step 1   Step 2   Step 3   Step 4

③   Explain why the expression you chose in Problem 2 is the best match.

_____

_____

_____

_____

Practice   Simplify.

④ $5(x + 10) =$ _____

⑤ $2(x + 8) =$ _____

⑥ $x + 2x =$ _____

Using Anthropometry

The following passage is from *Gulliver's Travels* by Jonathan Swift. The setting is Lilliput, a country where the people are only 6 inches tall.

"Two hundred seamstresses were employed to make me shirts . . . . The seamstresses took my measure as I lay on the ground, one standing at my neck, and another at my mid leg, with a strong cord extended, that each held by the end, while the third measured the length of the cord with a rule of an inch long. Then they measured my right thumb and desired no more; *for by a mathematical computation, that twice round the thumb is once round the wrist, and so on to the neck and the waist,* and by the help of my old shirt, which I displayed on the ground before them for a pattern, they fitted me exactly."

1. Four body parts are referenced in the text. What are they? Choose a variable to represent each one.

   _____   _____

   _____   _____

2. Take these four measures on yourself, measuring to the nearest $\frac{1}{4}$ inch.

   _____   _____   _____   _____

3. Use the variables you recorded in Problem 1 to write three rules described in the text.

   _____   _____   _____

4. Based on your data, how well do you think Gulliver's new clothes fit? Explain.

   _____

   _____

Practice

Evaluate.

5. $5\frac{1}{2} \div \frac{1}{4} =$ _____

6. _____ $= 2\frac{2}{3} \div \frac{3}{4}$

Which Would You Rather Have?

Don's boss is offering him two choices to get paid for June.

- Choice #1 is to receive $10 on June 1st, $20 on June 2nd, $30 on June 3rd, and so on through June 30th.
- Choice #2 is to receive 1 penny the first day, 2¢ the second day, 4¢ the third day, and so on, doubling the amount each day for the rest of the month.

① **a.** Predict which is the better plan. _____

b. Explain how you made your choice. _____

_____

② Enter formulas to complete the table for the first five days of each plan.

☐ **Boxes**					☒
C16 ▼ ⊙ *fx*					
	A	B	C	D	E
1	June Date	Choice 1	Choice 1 Total So Far	Choice 2	Choice 2 Total So Far
2	1	10.00	10.00	0.01	0.01
3	2				
4	3				
5	4				
6	5				

③ Use a spreadsheet program or a calculator to determine how much Don would receive for the day on June 30th for each choice.

Choice 1: _____   Choice 2: _____

④ If you have a spreadsheet program, find the total amount Don receives for both choices. If you do not, explain how to find the totals on the back of this page.

Choice 1: _____   Choice 2: _____

Practice Write three equivalent ratios for each ratio.

⑤ 2.5 to 2 _____   ⑥ 1 : 1.4 _____

⑦ $\frac{1}{2}$ to 3 _____   ⑧ $\frac{1}{2} : \frac{3}{4}$ _____

227

Congratulations!

By completing *Sixth Grade Everyday Mathematics*, your child has accomplished a great deal. Thank you for your support.

This Family Letter is intended as a resource for you to use throughout your child's vacation. It includes an extended list of Do-Anytime Activities, directions for games that you can play at home, a list of mathematics-related books to get from your local library, and a preview of what your child might be learning in seventh grade.

Do-Anytime Activities

Mathematics means more when it is rooted in real-world situations. To help your child review many of the concepts learned in sixth grade, we suggest the following activities for you to do with your child during vacation. These activities will help your child build on the skills that he or she has learned this year and are good preparation for a seventh-grade mathematics course.

1. Practice quick recall of multiplication facts. Include extended facts, such as $70 * 8 = 560$ and $70 * 80 = 5,600$.

2. Practice calculating mentally with percents. Use a variety of contexts, such as sales tax, discounts, and sports statistics.

3. Use measuring devices—rulers, metersticks, yardsticks, tape measures, thermometers, scales, and so on. Measure in both U.S. customary and metric units.

4. Estimate the answers to calculations, such as the bill at a restaurant or store, the distance to a particular place, miles per gallon on a trip, the number of people at an event, and so on.

5. Play games like those in the *Student Reference Book*.

6. If you are planning to paint or carpet a room, consider having your child measure and calculate the area. Have him or her write the formula for area ($A = l * w$) and then show you the calculations. If the room is an irregular shape, divide it into separate rectangular regions and have your child find the area of each one.

7. Ask your child to halve, double, or triple the amount of each ingredient in a particular recipe. Have your child explain how he or she calculated each amount.

8. Help your child use ratios in relation to the wins and losses of a favorite sports team. Ask him or her to decide which ratio is being used. For example, wins to losses (such as 5 to 15) or losses to wins (15 to 5) are part-to-part ratios. Part-to-whole ratios are used to compare wins to all games played (5 out of 20) or losses to all games played (15 out of 20).

9. Provide extra practice with partial-quotients division by having your child divide 3-digit numbers by 2-digit numbers, 4-digit numbers by 3-digit numbers, and so on. Ask your child to explain the steps of the algorithm to you as she or he works through them.

Building Skills through Games

The following section lists directions for games that can be played at home. Regular playing cards can be substituted for the number cards used in some games. Other cards can be made from index cards.

Name That Number

Materials	number cards 0–10 (4 of each) and 11–20 (1 of each)
Players	2 or 3
Skill	Naming numbers with expressions
Object of the Game	To collect the most cards

Directions

1. Shuffle the deck and deal five cards to each player. Place the remaining cards number-side down on the table between the players. Turn over the top card and place it beside the deck. This is the **target number** for the round.

2. Players try to match the target number by adding, subtracting, multiplying, or dividing the numbers on as many of their cards as possible. A card may only be used once.

3. Players write their solutions on a sheet of paper. When players have written their best solutions:

   • Each player sets aside the cards they used to match the target number.

   • Each player replaces the cards they set aside by drawing new cards from the top of the deck.

   • The old target number is placed on the bottom of the deck.

   • A new target number is turned over, and another round is played.

4. Play continues until there are not enough cards left to replace all the players' cards. The player who has set aside the most cards wins the game.

Getting to One

Materials	1 calculator
Players	2
Skill	Estimation
Object of the Game	To correctly guess a mystery number in as few tries as possible

Directions

1. Player 1 chooses a mystery number that is between 1 and 100.

2. Player 2 guesses the mystery number.

3. Player 1 uses a calculator to divide Player 2's guess by the mystery number. Player 1 then reads the answer in the calculator display. If the answer has more than 2 decimal places, only the first 2 decimal places are read.

> **NOTE** For a decimal number, the places to the right of the decimal point with digits in them are called *decimal places*. For example, 4.06 has two decimal places, 123.4 has one decimal place, and 0.780 has three decimal places.

4. Player 2 continues to guess until the calculator result is 1. Player 2 keeps track of the number of guesses. Player 2 may wish to keep track of guesses by recording them in a "What's My Rule?" table such as this:

in	out

5. When Player 2 has guessed the mystery number, players trade roles and follow Steps 1–4 again. The player who guesses his or her mystery number in the fewest number of guesses wins the round. The first player to win three rounds wins the game.

Vacation Reading with a Mathematical Twist

Books can contribute to learning by presenting mathematics in a combination of real-world and imaginary contexts. Teachers who use *Everyday Mathematics* in their classrooms recommend the titles listed below. Look for these titles at your local library or bookstore.

Problem-Solving Practice

Math for Smarty Pants by Marilyn Burns (Yolla Bolly Press, 1982)

Brain Busters! Mind-Stretching Puzzles in Math and Logic by Barry R. Clarke (Dover Publications, 2003)

Wacky Word Problems: Games and Activities That Make Math Easy and Fun by Lynette Long (John Wiley & Sons, Inc., 2005)

My Best Mathematical and Logic Puzzles by Martin Gardner (Dover Publications, 1994)

Math Logic Puzzles by Kurt Smith (Sterling Publishing Co., Inc., 1996)

Skill Maintenance

Delightful Decimals and Perfect Percents: Games and Activities That Make Math Easy and Fun by Lynette Long (John Wiley & Sons, Inc., 2003)

Dazzling Division: Games and Activities That Make Math Easy and Fun by Lynette Long (John Wiley & Sons, Inc., 2000)

Fun and Recreation

Mathamusements by Raymond Blum (Sterling Publishing Co., Inc., 1999)

Mathemagic by Raymond Blum (Sterling Publishing Co., Inc., 1991)

Kids' Book of Secret Codes, Signals, and Ciphers by E. A. Grant (Running Press, 1989)

The Seasons Sewn: A Year in Patchwork by Ann Whitford Paul (HMH Books for Young Readers, 1996)

Looking Ahead

Everyday Mathematics experiences in sixth grade prepare your child to do the following in future math classes:

- Use proportional reasoning to solve problems.

- Compute with fractions and decimals.

- Continue to write equivalent algebraic expressions to model and solve problems.

- Solve equations.

- Use formulas to solve problems.

Thank you for your support this year. Have fun continuing your child's mathematical experiences throughout the summer!

Best wishes for an enjoyable vacation.